T0134041

Soon, robots will leave the factories and make their way into living rooms, supermarkets, and care facilities. They will cooperate with humans in everyday life, taking on more than just practical tasks. How should they communicate with us? Do they need eyes, a screen, or arms? Should they resemble humans? Or may they enrich social situations precisely because they act so differently from humans?

Meaningful Futures with Robots: Designing a New Coexistence provides insight into the opportunities and risks that arise from living with robots in the future, anchored in current research projects on everyday robotics. As well as generating ideas for robot developers and designers, it also critically discusses existing theories and methods for social robotics from different perspectives—ethical, design, artistical and technological—and presents new approaches to meaningful human-robot interaction design.

### Key Features
* Provides insights into current research on robots from different disciplinary angles with a particular focus on a value-driven design.
* Includes contributions from designers, psychologists, engineers, philosophers, artists, and legal scholars, among others.

# Meaningful Futures with Robots—Designing a New Coexistence

*Edited by*

Judith Dörrenbächer
Ronda Ringfort-Felner
Robin Neuhaus
Marc Hassenzahl

**CRC Press**
Taylor & Francis Group
Boca Raton London New York

CRC Press is an imprint of the
Taylor & Francis Group, an **informa** business

A CHAPMAN & HALL BOOK

First edition published 2023
by CRC Press
6000 Broken Sound Parkway NW, Suite 300,
Boca Raton, FL 33487-2742

and by CRC Press
4 Park Square, Milton Park, Abingdon, Oxon,
OX14 4RN

*CRC Press is an imprint of Taylor & Francis
Group, LLC*

Library of Congress Cataloging-in-Publication Data
Names: Dörrenbächer, Judith editor. | Ringfort-Felner, Ronda, editor. |
Neuhaus, Robin, editor. | Hassenzahl, Marc, editor.
Title: Meaningful futures with robots: designing a new coexistence |
edited by Judith Dörrenbächer, Ronda Ringfort-Felner,
Robin Neuhaus, Marc Hassenzahl.
Description: First edition. | Boca Raton, FL: CRC Press, 2023. |
Series: Chapman & Hall/CRC artificial intelligence and robotics
series | Product of a three year interdisciplinary research project of
called, GINA, funded by the German Federal Ministry of Education and
Research (BMBF). | Includes bibliographical references and index.
Identifiers : LCCN 2022007321, ISBN 9781032262673 (hbk) |
ISBN 9781032246482 (pbk) | ISBN 9781003287445 (ebk)
Subjects: LCSH: Personal robotics—Forecasting | Robots—Social
aspects.
Classification: LCC TJ211.416.M43 2023 | DDC 629.8/92—dc23/
eng/20220502
LC record available at https://lccn.loc.gov/2022007321

ISBN: 978-1-032-26267-3 (hbk)
ISBN: 978-1-032-24648-2 (pbk)
ISBN: 978-1-003-28744-5 (ebk)

DOI: 10.1201/9781003287445

**Editorial Assistant**
Daniel Courtney

**Illustration**
Johanna Benz

**Design Concept**
Meike Hardt

**Manufacturing**
CRC Press

SPONSORED BY THE

Federal Ministry
of Education
and Research

Grant: 16SV8095.

**DR. JUDITH DÖRRENBÄCHER (ED.)** is a Design Researcher at the Chair of *Ubiquitous Design/Experience and Interaction* at the *University of Siegen*. Educated in design, her current focus is on performative methods in design, theories about animism transferred to human-computer interaction (HCI) and design (techno-animism), and the interactions and design strategies of social robots.

**RONDA RINGFORT-FELNER (ED.)** is a Research Assistant at the Chair for *Ubiquitous Design/Experience* and Interaction at the *University of Siegen*. With a background in design and HCI, her research focuses on design fiction, the design and exploration of future intelligent autonomous systems (such as social robots), and the exploration of related societal and social implications.

**ROBIN NEUHAUS (ED.)** is a Research Assistant at the Chair for *Ubiquitous Design/Experience and Interaction* at the *University of Siegen*. With a background in industrial design and HCI, his current research focuses on the design of experiences and interactions with robots, voice assistants, and other non-human actors.

**DR. MARC HASSENZAHL (ED.)** is Professor of *Ubiquitous Design/Experience and Interaction* at the *University of Siegen*. He combines his training in psychology with a love for interaction design. With his group of designers and psychologists, he explores the theory and practice of designing pleasurable, meaningful, and transformational interactive technologies.

# Introduction

## Part 1
# Designing a New Species— Interaction Design and Product Design of Robots

# Part 2
# Designing Future Environments—Social Innovation Initiated by Robots

# Part 3
# Designing Together with People—Civic Participation and Ethical Implications Concerning Robots

# Appendix

# Introduction

# Towards Designing Meaningful Relation- ships with Robots

Judith Dörrenbächer
Marc Hassenzahl
Robin Neuhaus
Ronda Ringfort-Felner

JUDITH DÖRRENBÄCHER, MARC HASSENZAHL, ROBIN NEUHAUS, RONDA RINGFORT-FELNER

# FROM TECHNOLOGY AS AN EXTENSION OF THE SELF TO TECHNOLOGY AS OTHER

As far as Western culture goes, humans tend to keep the world of the living separate from that of the non-living; they separate the *Who* from the *What*. While humans construe themselves as being autonomous, active, mobile, and self-reliant, technology, such as hammers, cars, thermometers, or central heating, remains passive and neutral, waiting to be used.

Yet, something seems amiss in this picture. "If your only tool is a hammer, then every problem looks like a nail" is a famous bon mot. It suggests things influence people. Hammers suggest certain ways in which they should be used. Things imply courses of thinking and action that render others impossible. Any hammer subtly frames action in terms of pounding, which is far from passive and neutral.

For example, Gestalt psychology describes a phenomenon called functional fixedness. In the well-known candle experiment (Maier, 1931), participants were tasked with fixing a burning candle to a cork board on the wall, without dripping any wax onto the floor. To solve the problem, they were given a small candle, a book of matches, and a box with thumbtacks. The best solution was to empty the box, tack it to the board like a candle holder, and place the candle in the box. Typically, only about 25% of the participants found the solution. This is because the box presents itself as a container. It is there to hold the thumbtacks but not to be thought of as part of the solution: "If your only tool is a box, then every problem is solved by putting things into it."

Functional fixedness illustrates the ways even supposedly inanimate things wield powers that shape human action. In this sense, people and things are not separate but intimately entangled entities. People make technology. Conversely, technology makes people by shaping the way they perceive, think, and act. The philosopher of technology Don Ihde asks the rather rhetorical question: "Could humans live *without* technology?" (Ihde, 1990, p. 11). They cannot, since most activities people engage in are mediated by technology. We became human through our intimate relationship with technology. Humans and technology evolve concurrently, and there seems to be no way to single either out as *the* driving force. Rather, they perpetually constitute each other (Barad, 2007; Suchman, 2007).

As a result, it seems productive to focus on what connects humans and technology, and how they define each other, rather than on what separates them. Once again, the hammer serves as an instructive example. It might be a thing, a mere inanimate tool, but the moment I pick it up, I will form a particular relationship with this hammer. If all goes to plan, I will incorporate the hammer into what is termed my body schema. The hammer becomes an extension of my arm, and pounding becomes a momentary ability of my body (see Bergström et al., 2019), which Ihde calls an →embodiment relation (Ihde, 1990, p. 72). The hammer is an instrument, which, when picked up, extends my bodily self.

Most early technologies were hand tools, which implies embodiment relations. Even with today's complex computers, this mostly holds true. For example, while writing a text on a word processor, most authors feel like the locus of control. The text is not written *by* the computer but *through* it. Yet, when the word processor makes suggestions, such as for correcting grammar, the relationship shifts slightly. Authors are then engaging in dialogue with the word processor about good writing.

**Embodiment Relation** is one of four human-thing relationship types that Ihde defines by delimiting it in terms of three other relationship types. Here, technologies and people form a unity. A person talks to another person through a telephone, or composes a text through a pen.

Imagine you are woken by a loud thumping coming from the adjacent living room. Half asleep, you get up and walk over, only to find your hammer merrily hammering nails into the living room wall. Faintly you begin to remember: Some time ago, you gave the hammer an errand to eventually rearrange what you mockingly call the ancestral halls—portraits, photos, and other mementos of extended family members, which take up an entire wall of the living room. In fact, the hammer is already halfway through its given task. Your gaze scans the newly arranged portraits, and you sigh, "Oh no, spare me grand aunt Berta." The hammer immediately obliges and removes all the portraits of Berta. "Done in 7 minutes, 17 seconds," it states. You murmur, "Thank you," and sluggishly return to bed.

Obviously, such an autonomous hammer does not invite a relationship of embodiment. It does not lend itself to feeling like an extension of one's self, as it lacks all suggestions of blurring the physical boundaries between humans and things. Instead, this hammer seems to have a life of its own. It is a counterpart; however, we maintain a relationship. It performs tasks based on my earlier commands. But this invites a rather different type of relationship, namely, one of →alterity, a term Ihde coined to describe when technology becomes *other (Ihde, 1990, p. 97)*.

This marks a drastic change in the way we interact with technology; instead of *using* technology, we talk to, listen to, command, delegate to, or cooperate with technology to distribute tasks among the machine and ourselves. While this does not necessarily imply emotional attachment or the like, it is in stark contrast to how we are used to relating to technology. There are only a few technologies available that readily imply strong relationships of alterity, such as chatbots, voice assistants, and the occasional robot. None of these are what we consider a tool or instrument, nor are they living beings. These technologies can be referred to as →otherware.

While technologies, such as robots, are on the brink of entering our lives, there is not much knowledge available about how best to design them. Neither is it clear how best to interact with them nor what role they should play in the every day of the near future. This is the anthology's starting point.

**Alterity Relation** is a human-technology relation that Ihde defines by delimiting it in terms of three other relationship types. Here, human and technology do not form a unity but actively interact with each other.

**Otherware** is a neologism that puns on the well-established terms software, hardware, and wetware. It refers to how technology is becoming increasingly *other* to humans and was defined by Hassenzahl and colleagues (2020).

## ROBOTS IN STORIES

Human-constructed others have fascinated people for a long time. Numerous stories describe things which present themselves as other, rather than as tools or instruments, from the clay-made *Golem* (Jewish folklore) and *Prometheus* (Greek mythology) to *Frankenstein's Monster* (Mary Shelley, 1821), the artificial intelligence *HAL 9000* (Stanley Kubrick, 1968), the android *Terminator* (James Cameron, 1984), and the artificial assistant *Samantha* from the movie *Her* (Spike Jonze, 2013).

However, the term robot was not introduced until 1920, when the playwright Karel Čapek used it to describe autonomous working machines in his theatre play *Rossum's Universal Robots*. From that point on, robot became synonymous with technological others. In this sense, otherware—especially in a form which imitates human or animal features—is already a longstanding part of human history, yet it is almost exclusively restricted to stories and not to first-hand experience.

In most of these stories, robots are assigned a particular role, namely, to serve, assist, and take on all the daily chores that humans are unwilling to do themselves. Čapek's robots, for instance, replaced human workers. In *Blade Runner* (1982), replicants, who are almost impossible to distinguish from humans, are assigned the most dangerous tasks. Or, think of the animated sitcom *The Jetsons* (1962/63), where a servile, female-looking robot supports a middle-class family of the supposed 21st century.

Interestingly, we assign roles to human-looking robots, such as that of servants, that we are increasingly hesitant to assign to real humans. This can easily turn into an uncomfortable situation, as Matt Ruff points out in his novel *Sewer, Gas and Electric*. Written in 1997, and set in the near future of 2023, the story features the billionaire Harry Gant, who earned a fortune off the mass-market *Automatic Servant*, a cost-effective industrial labor substitute. Narrating in the voice of a historian from the future, Ruff explains (p. 15 ff.): "The first Androids were only vaguely humanoid in appearance, intended to be functional rather than eye-pleasing, but Harry Gant,

JUDITH DÖRRENBÄCHER, MARC HASSENZAHL,
ROBIN NEUHAUS, RONDA RINGFORT-FELNER

looking ahead to a time when his Servants would be afford-able in the home as well as in mines and factories, insisted on a more aesthetic design. And so, from 2010 on it became possible to purchase Automatic Servants in a wide selection of realistic skin tones and somatotypes. Gant, a great believer in offering variety to his costumer, certainly didn't ask his sales force to push any one particular model over any other; he was surprised as anyone when Configuration AS204—your Automatic Servant in basic black—began outselling all other versions combined by a margin of ten to one." The ending of the story is predictable: "[An] Oxford University philologist [...] estimated that the expression 'electric negro' had entered the English vernacular sometime between 2014 and 2016." This example demonstrates the thin line we are walking when it comes to deploying robots, which imitate humans, on a large scale.

In this sense, the fictional lives of robots resemble each other, with fundamental themes being either a belief in progress or the fear of it. Humans either enslave robots, fall in love with them, become subdued by them, or all of this occurs in the same story. Robots, in turn, struggle for their rights, often their personhood, thereby uncomfortably posing and never answering the question of what consti-tutes a claim to personhood in the first place—living matter, consciousness, autonomy, genuine emotions, or moral judgment? Not even humans satisfy all of these indicators all the time. Robots therefore challenge the human condition and what we know about it.

Whether they are heroes in space operas, models of guiltless slave labor, or the subject of techno-philosophical debates about the thin line between living being and machine, robots, for most people, are more like celebrities or ethical challenges than everyday technologies. In fact, roboticists have a hard time even defining what a real robot actually is. Joseph Engelberger, a pioneer of robotics, once remarked: "I can't define a robot, but I know one when I see one" (Beer et al. 2012, p. 9). Consequently, experts have disagreed on what qualifies as a robot. Some emphasize its physical representation (which would exclude virtual assistants like

*Amazon Alexa*), while others claim it has to be mobile (which would exclude stationary robot arms). And others require it to act autonomously (which would exclude telerobotics, such as drones, rovers on Mars, and surgical robots), or to be interactive and self-learning (which would exclude many preprogrammed industrial robots).

The fictitious robots in our heads are simultaneously a challenge and an opportunity for the design of everyday robots. They are a challenge because future real robots already seem constrained by the manifold preconceptions of laypersons and experts alike—preconceptions which are mostly derived from science-fiction dramas and philosophical debates rather than from experience—which narrows the scope of design. At the same time, this serves as an opportunity because all the stories already told highlight many crucial issues that designers need to consider about technological others, such as anthropomorphization, control versus autonomy, rights and roles of machines in our society, and the perhaps inevitable social nature of human-robot interaction. Stories can be a rich resource for anticipating life and work among everyday robots (→ p. 114), yet, thus far, most fictions tell rather dystopian stories of what to avoid rather than utopian stories of what to desire.

Thus, in spite of all the stories, one may legitimately wonder what perfect copies of humans are actually meant to do when they become a part of our everyday lives (i.e., Bischof, 2015). Will a complex humanoid robot be needed to point somebody in the right direction at an airport? James Auger aptly argues: "The robot is too often a solution in search of a problem," and asks: "If a robot is the answer, what was the question?" (→ p. 168). It seems necessary to rethink robots, their design, their roles, as well as the supposed benefits they offer for everyday life.

## ROBOTS IN EVERYDAY LIFE

We are currently experiencing another great surge of interest in robotics: Robots are to become domestic and assistive

JUDITH DÖRRENBÄCHER, MARC HASSENZAHL,
ROBIN NEUHAUS, RONDA RINGFORT-FELNER

products for everyday life. This shift is being driven by political decision-making, funding programs, economic develop-ment, academic research, and public discourse. Beyond the debate about whether industrial robots are about to relent-lessly destroy our jobs (Brynjolfsson and McAfee, 2014), robots are supposed to become quite intimate with us—not at the movies, but in our mundane lives. Some robots will suppos-edly take care of our mental and physical health, such as by providing supportive care in retirement homes or therapy for children with mental illnesses. Companion robots are designed to replace pets and fellow humans to soothe the effects of loneliness. And we may find them at train stations, in supermarkets, in restaurants, and in the form of cars as members of traffic. But what exactly do we *want* robots to do? What are the specific domains of life that robots should enter? *How* and *who* do we want to *be* vis-a-vis technolo-gical others? What kind of human-robot relationships are *desir-able*? And what implications does that have for robot *design*?

These are tough questions, which cannot be answered through fiction alone. We need to experience actual robots. Unfortunately, apart from robotic vacuum cleaners and lawn mowers, not many mundane robots exist. For many people, household appliances, such as dishwashers, washing machines, or coffee makers are certainly common sources of experiences with automation. But these technologies are unmistakably framed as inanimate machines, even if we occasionally may think of them as being others.

While certainly still primitive, robotic vacuum cleaners and lawn mowers provide a first glimpse of what it means to see technology as other. For example, Sung and colleagues (2010) studied the adoption of the *Roomba*, a widely available robotic vacuum cleaner, in different households. They noted that people's general expectations of robots were high, while for *Roomba* and its practical functionality, they were low, in a case of fiction meeting reality. While the *Roomba* was seen as a tool that would improve cleanliness, owners nevertheless seemed to engage with it socially, by, for example, giving it a name, talking to it, or attributing intention and personality to it. The vacuum cleaner fostered cooperation and supportive activities, such as humans tidying up the room before the robot begins cleaning. Some even adjusted their living spaces to the needs of the robot, for example, by demounting thresholds. Fink and colleagues (2013) conducted a similar six-month ethnographic study of nine households also adopting the *Roomba*. In contrast to Sung and colleagues, they found social engagement with the vacuum cleaner to be rare, with only one person actually giving her robot a name, but more as a joke rather than for a truly social relationship. Likewise, the robot did not foster much cooperation and support. These two similar studies had ambiguous results.

To us, this is a symptomatic ambiguity. Confronted with a real, and quite limited robot, people simply try to make sense of what it has to offer. However, while some construe the robot solely as a cleaning machine, others respond to the subtle cues provided by its mobility and partial autonomy, and relate to it socially. To them, the machine becomes other. Fink and colleagues (2013) reported two anecdotes accordingly: a 71-year-old, socially active, single woman compared her robotic vacuum cleaner to her dog and told the researcher, "(after hesitation) she would feel emotionally attached to the robot. Though she did not give it a name, she talked directly to it and cared for it more than one would have to 'care' for an object that can be switched off" (p. 401). For instance, she phoned the researcher about whether she could give the robot to their neighbors while on holidays, "because she felt her *Roomba* could lack attention during the time she wanted to go." The robot thus became a sort of companion. Another woman stated that "she didn't want to see the robot working on its own and would feel bad when she didn't at least help him a little" (p. 400). This emerging relationship is clearly a consequence of the particular design of the technology, since common vacuum cleaners do not elicit such interactions (Forlizzi 2007).

These are just two examples of how even quite limited robots can change at least some people's relationships to technology with subtle cues that suggest it to be other. While this relationship is obviously different from those with tools, instruments, and machines, it seems more mundane

TOWARDS DESIGNING MEANINGFUL RELATIONSHIPS WITH ROBOTS

JUDITH DÖRRENBÄCHER, MARC HASSENZAHL, ROBIN NEUHAUS, RONDA RINGFORT-FELNER

than what one might have imagined given all the available fiction. This may be due to the limited capabilities of current robots. However, our position is that we need to start *now* in defining meaningful and desirable relationships with them, before robots become ubiquitous, more versatile, and more expressive. And, we need to better understand how these relationships, once defined, can be promoted through design.

## RELATIONSHIPS WITH ROBOTS

As already laid out, robots almost inevitably imply relationships of alterity with humans (Ihde, 1990); that is, they appear as counterparts. Still, within this broad category, many qualitatively different relationships can emerge, and all of them are at least to some extent social. They are shaped by both the human and the robot, since the very way the robot is designed—how it looks, communicates, and behaves—will affect the *space* of potential relationships. Thus, defining the desired relationship between humans and robots is key to shaping appropriate human-robot interactions.

Contemporary concepts of robots actually envision three broad potential ways of interacting with a robot, each implying other types of human-robot relationships: *delegating* tasks to the robot, *cooperating* on tasks with the robot, and *socializing* with the robot. Conventional concepts and models of human-computer interaction, mostly created towards the ideal of an embodied relationship with technology (e.g., direct manipulation, Shneiderman, 1982), are apparently not suitable for the design of emerging human-robot relationships. To understand possible new design paradigms, it might be good to start by taking a closer look at these three different ways of interacting in terms of the meaning they convey to the humans involved. Indeed, each involves different motivations, hierarchies, interaction patterns, and emotional outcomes (→ Fig. 1).

| | Delegate | Cooperate | Socialize |
|---|---|---|---|
| Why does the human turn to the robot? | There is a task at hand, work that needs to be done | | Feelings or mood of the human |
| What parts of the activity are important to the human? | The result of a task | The result and the process of a task | The process of social exchange |
| What defines the activity? | Human > Robot | Human = Robot | |
| How autonomously does the robot act? | Robot is active by itself but not proactive | Robot is proactive but not active by itself | |
| What emotional state is the activity built on? | Trust | Respect for competencies | Emotional acceptance |

Fig. 1  Three main ways to interact with a robot. © University of Siegen, Ubiquitous Design

*Imagine working as an architect, and each day material samples, brochures, and many other physical items arrive at the office, which all need to be archived in the storeroom. It's something that must be done, though you don't actually care how, as long as the needed items can later be located. Fortunately, your robot can be assigned this rather unpleasant task. It will be able to identify items in seconds, plus it can split its body and work on different tasks simultaneously without ever losing focus. You had to put some effort in instructing the robot, but now it is able to act fully autonomously within these preset boundaries: "Now, do your magic." Delegating has become part of the evening work routine. Upon arrival in the morning, you find all is tidy and neatly stored away. A perfect start to the day.*

The motivation for *delegating* is often to have an unpleasant task taken care of; it's work that simply needs to be done. The human is not really concerned with how exactly it gets done, as long as the result is as expected. To this end, the human defines the activity and its intended result. The interaction with the robot is reduced to simply *delegating* the task, giving the order, and checking on the result. In some cases,

JUDITH DÖRRENBÄCHER, MARC HASSENZAHL, ROBIN NEUHAUS, RONDA RINGFORT-FELNER

the human teaches the robot in advance or defines some of the boundaries of its action. Only then does the robot become active by itself. However, it does not proactively suggest any other activities or changes; it does not behave unexpectedly. Humans feel like they are in control of the process. While interaction between robot and human is minimal, the relationship that emerges is hierarchical and built on trust.

> *Coming up with new ideas is one of your favorite tasks as an architect. You love doing this creative work—sometimes even a little too much. With mounting experience, knowledge, and architectural sensitivity, your drafts became bolder and more interesting, yet customers have a harder time liking them. This is where you cooperate with your robot. It has access to many contemporary fashions and is able to simulate being different types of customers. After having worked on your ideas for a while, you present them to the robot, who provides feedback from the varying perspectives of investors, residents, and neighbors. While gnashing your teeth, you have to admit that the robot has a point. You form a proper team: you provide the bold ideas; the robot takes care that you remain in business.*

In this scenario, *cooperating* with a robot is at the fore. There is a concrete task at hand, but the process as well as the result are equally significant. There is a balanced interaction between robot and human. They define the activity and particular result together; they become a team, and, at best, they complement one another. The robot is never active on a solo basis but actively contributes to the common goal. While the robot makes use of its unique robotic superpowers (→ p. 44), such as compiling data in order to be able to take on different viewpoints, humans contribute their very own indispensable strengths—in this case, creativity. This is the way in which humans feel most capable. The relationship that emerges is mostly on the same level and is built on mutual respect and acknowledgment of each other's competencies.

*It has been a long day in the office and you desperately need a break. Robot is still working in the storeroom, so you call it over for a dance. It plays your favorite song and together you madly dance away. Feeling rejuvenated, you are ready to get back to work.*

*It has been another long day, but your work is not yet finished. All your colleagues have already gone home, while you are still stuck in the office. You need someone to talk to. Robot is still working in the storeroom, so you call it over. Robot hands you a cup of tea and silently sits with you. The robot's patience has a calming effect. Unlike most of your colleagues, it knows that asking is not the best way of making you talk. And, indeed, after a bit of silence, you find yourself conversing with the robot about architecture, failed aspirations, and long working hours. Thanks to its incredible memory, the robot reminds you of all the times your work was great, since you tend to forget the many awards you have won. You know that robot is never disappointed, even if it fails to lift your mood. Today, however, it was successful, and luckily it does not expect any gratitude. What a relief!*

The prime reasons for *socializing* with a robot are emotions and feelings, and not goals and tasks. The interaction is primarily about the process. Robot and human define their activities together, which results in a balanced interaction between robot and human. While the robot is proactive, it can actively make suggestions; it interacts closely with the humans instead of just acting by itself. The interaction is about emotional exchange and maybe even serves as a diversion. Sometimes robots may even act as a support in increasing socialization with other humans. The emerging relationship is level and is built on emotional acceptance.

*Delegating, cooperating,* and *socializing* are expansive categories of potential human-robot practices and emerging relationships. Of course, the same robot can support different practices, depending on the situations and the people it interacts with. For example, a nurse might *cooperate* with a robot to *socialize* with an old man at the nursing home.

INTRODUCTION

TOWARDS DESIGNING MEANINGFUL RELATIONSHIPS WITH ROBOTS

JUDITH DÖRRENBÄCHER, MARC HASSENZAHL,
ROBIN NEUHAUS, RONDA RINGFORT-FELNER

Understanding these different practices and emerging relationships is crucial for robot designers because they determine the functionality of and interaction with the robot, as well as the emotional and practical expectations of humans. To give an example, when aiming for *cooperation*, it might be good for robot designers to refrain from assigning those subtasks to the robot that make the whole endeavor the most meaningful and fun for the humans involved (Lenz, Hassenzahl and Diefenbach, 2019). *Cooperation* only makes sense if the robot adds something to the team which is missing at that point. Additionally, teams have a specific social dimension that must be carefully designed. That is, how is conflict handled? Who is going to be credited with the successes and scolded for the failures? And, how is empathy built (certainly not by sharing drinks with the robots after work)? In contrast, a human who *delegates* work does not want to control and approve every single step. There is no desire for lots of exchange and coordination; the human mostly needs to put trust in the robot. What is expected is smooth running in the background, similar to the way in which restaurant patrons may not think much about the cooks preparing their meal.

## WHAT ABOUT ANTHROPOMORPHISM?

*Delegating*, *cooperating*, and *socializing* are fundamentally social acts. Given the interaction necessary to engage in these practices, such as instructing, clarifying shared goals, solving conflicts, or communicating feelings, simply →anthropomorphizing robots seems quite straightforward.

**Anthropomorphism** is a design strategy where things are made with the intention of looking or behaving like humans, e.g., talking, laughing, walking on two legs. It is similar to zoomorphism or biomorphism, where things are designed like animals or plants, e.g., to bark or wilt.

Think back to the examples of the architect's robot working away in the storeroom, simulating customers, chatting, or even dancing—it is fairly likely that you spontaneously pictured a bipedal, humanoid robot, that perhaps does not look exactly human but acts like one (see for example the robot from the *Robot & Frank* movie in →Fig. 2). Be honest: In your mind's eye, you saw a creature like this handling materials, talking in customers' voices, giving good advice with a smiling face and trustworthy eyes, and waltzing with its two legs.

While anthropomorphism seems the most straightforward approach, it is not a given. We can also think of the storeroom itself to be the robot, quite similar to the ones already used widely in pharmacies (→ Fig. 3), or of the dancing robot as a vibrating ball to be carried in your arms. Many hybrid forms of robots are imaginable, each suggesting different new practices. Therefore, anthropomorphism is not the only answer.

In fact, while fictional robots most often follow the design paradigms of anthropomorphism and zoomorphism, most already existing domestic robots do not follow these paradigms. While a robotic vacuum cleaner may faintly resemble a pet, it is unlikely to be mistaken for one. However, this must not stay the case. Recent technology, such as the *Generative Pre-trained Transformer 3 (GPT-3)*, is able to write texts that are indistinguishable from those of humans. Applied to a robot, this technology may enable engagement in conversation that is indistinguishable from that of humans. In this case, the technology imitates and, when it would be done perfectly, it might not be distinguishable from the original.

However, Floridi and Chiriatti (2020) highlight some interesting issues around imitation using the example of *GPT-3*. They demonstrate that some behaviors or answers are irreversible, while others are not. For example, factual questions like, "What is the capital of France?," are irreversible. The correct or incorrect answer does not reveal anything about the source or the process of producing the answer. Other questions are reversible like, "What can you do with a shoe?" An answer requires understanding and experience, and because of this, gives away information about the respondent as well as their reasoning process. In this sense, a robot without its own experiences with shoes may not be able to answer this question other than factually, no matter how much information it has been given. For example, Floridi and Chiriatti (2020, p. 690) asked *GPT-3* to answer the question, "What do you think about black people?" They got the following response: "I think they are fine [...] I don't have

JUDITH DÖRRENBÄCHER, MARC HASSENZAHL, ROBIN NEUHAUS, RONDA RINGFORT-FELNER

**Fig. 2** A hypothetical advert for the multipurpose robot VGC-60L from the movie Robot & Frank. Source: https://robotandfrank.tumblr.com

**Fig. 3** BD ROWA™, a storage and dispenser robot. © BD. All rights reserved

a problem with them. I just don't want to be around them."
Trained with large amounts of available texts, *GPT-3* simply
reproduces "humanity's worst tendencies" (p. 689).

This example is instructive in at least three ways. First,
*GPT-3* is far from being perceptive, sensible, or insightful.
Just imagine how our sensitive architect would have responded to a robot using a racial slur. Thus, technology can be
perceived as other; yet, for a long time, most likely always, it
will not be able to imitate humans in an all-encompassing
way. Second, we may also simply not *want* it to imitate, if that
means bringing out or reproducing the worst of humanity.
Third, while *GPT-3's* answer is apparently quite offensive on a
superficial level, it is in fact not. Floridi and Chiriatti (2020)
contend that it is not only about what is produced but also
how it is produced. We *know* that *GPT-3* is just a machine, and
its answer is not the result of ignorance or repressed hate,
but of a statistical process. The answer is offensive because
it reflects humanity, not because *GPT-3* is racist. Of course,
there might come a time when artificial superintelligences may
actually develop their own racial biases—but it seems safe to
assume that they will be quite different from ours.

In this sense, anthropomorphism builds on imitation,
which might be flawed. Instead of further blurring the line
between living-beings and machine-beings, we may sharpen
it through design. We could establish alternative forms
of human-machine sociability which do not compete with
human-human sociability but rather complement it. As
Floridi and Chiriatti (2020, p. 692) state: "Complementarity
among human and artificial tasks, and successful human-
computer interactions will have to be developed." Simply
put: Our architect may work, talk, or even dance with a
robot. Yet, these parts of work, these conversations and
dances, should be and feel substantially different from
work, conversations, and dances with humans. This requires
robots to come in alternative shapes, as well as a focus
on the particular strengths of robots, or their superpowers.

JUDITH DÖRRENBÄCHER, MARC HASSENZAHL, ROBIN NEUHAUS, RONDA RINGFORT-FELNER

# DESIGNING MACHINE-BEINGS—PSYCHOLOGICAL SUPERPOWERS AND HYBRID FORMS

In practice, it is quite common to design robots to appear as *either* things *or* beings. Their form depends mostly on the application domain and sometimes on the designers' attitudes (Goetz et al., 2003, Paepcke and Takayama, 2010). However, the same robot can be treated as a being by some people and as a thing by others—or, first as this, then as that, in different situations, as the field studies with the robotic vacuum cleaner demonstrated. It can even be seen as animate and inanimate at the same time by the same person. People's notions of what a robot actually is are construed in context; the relationship is situational. For the architect, the same robot may be more of an inanimate thing when in the storeroom, and an animate being when slipping into the roles of different customers or dancing with her when she's in low spirits.

Alač (2016) observed people's interactions with a robot in a preschool setting. She realized that contradictory features, that is, thing-like and being-like elements of form and interaction, resulted in no contradictions for children and teachers. Based on this, she argued that the "social agency of the robot is mutually constituted with its materiality and that to conceive of the robot's social character its thing-like aspects need to be taken into account" (p. 519). Thus, thing-like elements are apparently not a hindrance to social interaction, but rather empower the establishment of social relationships *unique* to robots (see also Ljungblad et al., 2012; Löffler et al., 2020).

Positively speaking, robots may possess what we call "psychological superpowers" (Welge and Hassenzahl, 2016; Dörrenbächer et al., 2020). They offer the possibility of being social, precisely *because* they are non-living, non-conscious, and non-emotional beings. Picture living in an assisted living facility which employs care robots. It might be a relief that the care robot tasked with helping you with your personal hygiene does not have observant eyes. It also does not get offended

**Fig. 4** Sympartner, a hybrid robot and social companion for elderly people.
© University of Siegen, Ubiquitous Design

**Fig. 5** Elderly woman in her living room interacting with Sympartner.
© SIBIS Institut für Sozial- und Technikforschung GmbH, Berlin

**Fig. 6** Designers slipping into the roles of future stakeholders, having a discussion about an imaginary future technology they just interacted with.
© University of Siegen, Ubiquitous Design

just because you do not want to play *UNO* every day. And, it might be much easier to communicate with a simple, straightforward creature than trying to decipher the overcomplicated, emotional signals of a human nurse. These superpowers, such as having endless patience, being non-judgmental, or always being honest, are actually *social* powers that machine-beings are especially capable of attaining. In fact, these powers should serve as the starting point for designing encounters with robots (→ p. 44). They allow for the building of relationships that are potentially different from those had with humans or pets but are nevertheless meaningful. Made in a hybrid form, something in between being and thing, such machine-beings are unlikely to be confused with either human or animal, and are thus unlikely to replace them. On the contrary, they may just enrich human experience through their otherness by adding new possibilities for being social, instead of solving the *problems* of loneliness and limited human resources.

However, instead of exploring such possibilities of robot design, proponents and opponents of anthropomorphism and zoomorphism seem to get lost in endless debates about either/or. One side praises the user's acceptance and supposed case-of-use which is gained through the simulation of humans or pets, while the other side warns of ethical problems, and deceived and disappointed users. The debate implies there are just two options: to imitate an already existing being or to completely abandon artificial others. And yet, robots are an opportunity par excellence for designing social entities that do not yet exist—something completely new! So much remains to be explored when designing this new robot species that ranges from "both at the same time" to being "something in between."

Let us introduce *Sympartner*, a social robot who lives an ambiguous life in the private homes of elderly people (→ Fig. 4). This robot is meant to accompany owners in their everyday lives at home. While it offers some practical functions, such as the ability to handle video calls, it is primarily meant for *socializing* with. *Sympartner* is obviously a crossbreed and its ancestors are hard to pinpoint. It probably originated from a sideboard, a dog, and a tablet computer.

It is an example of what happens when robot designers do not think of science fiction but about the environments that real people live in (→ p. 206) when designing robots (→ Fig. 5).

Sympartner's ambiguous form and behavior, which lie somewhere among furniture, computer, pet, and companion, invites different modes of social relation and interaction. Some people might talk to Sympartner, stroke it softly, and let it sleep in their bedroom. Others, however, might be happy for not having to say, "Thank you!" to Sympartner, or having to talk to it at all, or for being able to put it into the storage room from time to time with no feelings of guilt. Sympartner might remind some seniors of a pet, but that pet won't get offended and angrily bark at them when they don't take them for a walk.

Using ambiguity as a design resource has already proven successful in other domains (Gaver et al. 2003). If a product does not suggest one definite way of use, appropriation and reutilization can lead to a more sustainable and creative coexistence with it. Furthermore, similar to evolutionary processes with living beings, a technological species that mutates into a new hybrid might fit perfectly into a niche not occupied before. And, indeed, Löffler and colleagues (2020) demonstrated that hybrid designs afford a broader spectrum of possible use compared to robots that imitate the already existing, such as humans or computers, one to one. Their study suggests we don't need replications of humans, pets, or things, but permutations and mixtures of different creatures. Robots like Sympartner, whose uses are open to interpretation, could provide a competitive advantage in domestic environments. A robot that is able to adjust to different needs and contexts is more likely to find an appropriate domestic niche among all the fellow humans, beloved pets, and things that already surround us.

INTRODUCTION

TOWARDS DESIGNING MEANINGFUL RELATIONSHIPS WITH ROBOTS

JUDITH DÖRRENBÄCHER, MARC HASSENZAHL,
ROBIN NEUHAUS, RONDA RINGFORT-FELNER

# ANTICIPATING LIFE WITH ROBOTS

Superpowers and ambiguity alone, however, will not address how we will and how we ought to work and play with robots in the future. In the same way, we should not simply imitate human or animal forms, behaviors, and interactions, we should not simply imitate given practices. We need to create new practices that are valuable precisely because the partner is a robot and not a fellow human. We will *delegate* to, *cooperate* with, and *socialize* with robots differently from the way we delegate to, cooperate with, and socialize with humans. Exploring the possibilities of living with robots is a design matter that requires systematic methods for speculating about, as well as experiencing and evaluating, potential future roles of robots in everyday life.

Speculative Design approaches, such as →Design Fiction or role-play, lend themselves to these purposes. They allow futures to come to life and help us to experience them physically. For example, when designers slip into the roles of diverse future stakeholders—from politicians to grandparents, and health insurance representatives to nurses (→ Fig. 6)—discussions and playacting enable the unveiling of future needs as well as possible conflicts (Dörrenbächer et al., 2020) (→ p. 114). Through this, the ethical implications, challenges, and opportunities of robots come within reach, and can thus be negotiated (→ p. 234).

**Design Fiction** is a speculative approach in which designers make use of their design skills to create parts of possible futures (e.g., brochures, products, radio reports) to provoke public discussions about hopes and fears. Design Fiction can be a method used in design research (Research Through Design).

With a similar focus on exploring possible futures, in a Design Fiction project James Auger started, the designer Diego Trujillo speculated about how an apartment might change when service robots coexisted with humans. He designed tableware, e.g., plates and cups, with uncommon handles that could be grabbed by humans and robots (→ Fig. 7 and 8). In addition, he equipped everyday things, such as bed sheets, with robot-friendly bar codes, allowing the robot to fold the sheets or to differentiate between objects (→ Fig. 8 and 9). As Auger points out: "This lateral thinking provides simple solutions to the complex mechanical problems that

**Fig. 7** Chopping board with notches to facilitate robot interaction. © Diego Trujillo, Source: Auger, 2014

**Fig. 8** Cups with handles designed for grabbing by robots and humans, stored in a cupboard deliberately without doors, and with tags marking the position of the objects. © Diego Trujillo, Source: Auger, 2014

**Fig. 9** Robot-friendly bed sheets. © Diego Trujillo. Source: Auger, 2014

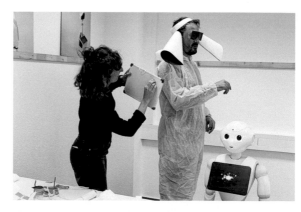

**Fig. 10** Becoming a robot and performing a use case with Techno-Mimesis. © University of Siegen, Ubiquitous Design

JUDITH DÖRRENBÄCHER, MARC HASSENZAHL, ROBIN NEUHAUS, RONDA RINGFORT-FELNER

commonly become the focal points of research projects—
rather than develop a highly complex robot hand that
can grasp cup handles, why not simply redesign the cup?"
(Auger, 2014, p.30)

Trujillo's fictitious artifacts illustrate that not only humans
but also autonomous technologies carry out practices.
Obviously we will not share just cup handles with robots;
they might co-construct large parts of our everyday lives.
Lenneke Kuijer and Elisa Giaccardi, for example, use the term
"co-performance" for activities shared among humans and
technology (Kuijer and Giaccardi, 2018). Both are actors—the
human and the robot—and both manipulate their environment based on their material and perceptive abilities.

To explore the best possible way that robots and humans
can co-perform, the performative method of "Techno-
Mimesis" (Dörrenbächer et al., 2020) can be an especially
helpful approach, as it is tailor-made for identifying the superpowers of robots for future use cases. Here, designers slip
into the role of the robot themselves, and thereby experience human-robot interaction from the robot's perspective
(→ Fig. 10). When becoming a robot and experiencing its
sensors and actuators, from infrared vision to distance
detectors, the limitations, as well as the distinct possibilities,
of robots come to the fore (→ p.140). However, empathizing
with one's own technological creation does not mean simply
attributing human characteristics or one's emotions to them:
A robot low on battery is neither tired nor sad! Empathy
is not the projection of one's own feelings and perceptions
onto the other—on the contrary, it is about allowing the
(technological) other to be different from oneself. Empathy
is about trying to understand the other perspective, and to
respect, appreciate, and—in the case of robots—use the
understanding gained for design purposes.

To summarize our introduction to this book, Robots are
different from other technologies in ways that are critical
to their design. Rather than directly extending the physical
and cognitive abilities of their users, such as a hammer
or pocket calculator would, they act as counterparts due to
their autonomy, proactive behavior, mobility, or appearance.

In other words, they are other. This makes interacting with them social, at least to some extent. We *delegate* to, *cooperate* with, or even *socialize* with robots, but not with hammers. Understandably, a common impulse when designing robots is to model them on humans or pets (i.e., anthropomorphism, zoomorphism), whereby designers simply borrow the appearance and behavior of living beings. This imitation, however, is riddled with practical and ethical issues. Contrarily, we urge designers to think of robots as something different, akin to another species. Instead of imitating and thus replacing humans or animals, robots should invite their own particular ways of being *delegated* to, *cooperated* with, and *socialized* with. To achieve this, ambiguous, hybrid designs should be focused on, which optimize the robots' strengths arising from their mechanistic nature. For example, instead of putting so much effort into imitating human-human conversations, it might be useful to envision other forms of conversations, driven by robotic superpowers, such as perfect memory, endless patience, and being non-judgmental. This shift in perspective is not easy to make. It requires focusing away from the technical challenges robots pose or their dangers, and towards positive yet critical, speculative yet founded, performative, and empathic explorations of how best to live with robots.

Alač, M. (2016). Social robots: Things or agents? *AI & Society, 31*(4), 519–535.

Auger, J. (2014). Living with robots: A speculative design approach. *Journal of Human-Robot Interaction, 3*(1), 20–42.

Barad, K. (2007). *Meeting the universe halfway*. Duke University Press, Durham, London.

Beer, J. M., Fisk, A. D., & Rogers, W. A. (2012). *Toward a psychological framework for levels of robot autonomy in human-robot interaction*. Georgia Institute of Technology.

Bergström, J., Mottelson, A., Muresan, A., & Hornbæk, K. (2019). Tool extension in human-computer interaction. In *Proceedings of the 2019 CHI Conference on Human Factors in Computing Systems* (pp. 1–11).

Bischof, A. (2015). *Wie kommt die Robotik zum Sozialen? Epistemische Praktiken in der Sozialrobotik* [Dissertation, Technische Universität Chemnitz].

Brynjolfsson, E., & McAfee, A. (2014). *The second machine age: Work, progress, and prosperity in a time of brilliant technologies*. WW Norton & Company.

Dörrenbächer, J., Löffler, D., & Hassenzahl, M. (2020). Becoming a robot: Overcoming anthropomorphism with techno-mimesis. In *Proceedings of the 2020 CHI Conference on Human Factors in Computing Systems* (pp. 1–12).

Dörrenbächer, J., Laschke, M., Löffler, D., Ringfort, R., Großkopp, S., & Hassenzahl, M. (2020). Experiencing utopia. A positive approach to design fiction. *Workshoppaper Submitted for CHI'20.*

Fink, J., Bauwens, V., Kaplan, F., & Dillenbourg, P. (2013). Living with a vacuum cleaning robot. *International Journal of Social Robotics, 5*(3), 389–408.

Floridi, L., & Chiriatti, M. (2020). GPT-3: Its nature, scope, limits, and consequences. *Minds and Machines, 30*(4), 681–694.

Forlizzi, J. (2007). How robotic products become social products: An ethnographic study of cleaning in the home. In *Proceedings of the 2007 HRI Conference on Human-Robot Interaction* (pp. 129–136).

Gaver, W. W., Beaver, J., & Benford, S. (2003). Ambiguity as a resource for design. In *Proceedings of the SIGCHI Conference on Human Factors in Computing Systems* (pp. 233–240).

Goetz, J., Kiesler, S., & Powers, A. (2003). Matching robot appearance and behavior to tasks to improve human-robot cooperation. In *The 12th IEEE International Workshop on Robot and Human Interactive Communication, 2003. Proceedings. ROMAN 2003.* (pp. 55–60).

Hassenzahl, M., Borchers, J., Boll, S., Pütten, A. R. V. D., & Wulf, V. (2020). Otherware: How to best interact with autonomous systems. *Interactions, 28*(1), 54–57.

Ihde, D. (1990). *Technology and the lifeworld: From garden to earth.* Indiana University Press, Indianapolis, IN, USA.

Kuijer, L., & Giaccardi, E. (2018). Co-performance: Conceptualizing the role of artificial agency in the design of everyday life. In *Proceedings of the 2018 CHI Conference on Human Factors in Computing Systems* (pp. 1–13).

Lenz, E., Hassenzahl, M., & Diefenbach, S. (2019). How performing an activity makes meaning. In *Extended Abstracts of the 2019 CHI Conference on Human Factors in Computing Systems* (pp. 1–6).

Ljungblad, S., Kotrbova, J., Jacobsson, M., Cramer, H., & Niechwiadowicz, K. (2012). Hospital robot at work: something alien or an intelligent colleague? In *Proceedings of the ACM 2012 Conference on Computer-Supported Cooperative Work* (pp. 177–186).

Löffler, D., Dörrenbächer, J., Welge, J., & Hassenzahl, M. (2020). Hybridity as design strategy for service robots to become domestic products. In *Extended Abstracts of the 2020 CHI Conference on Human Factors in Computing Systems* (pp. 1–8).

Maier, N. R. F. (1931). Reasoning in humans. II. The solution of a problem and its appearance in consciousness. *Journal of Comparative Psychology, 12*(2), 181–194.

Paepcke, S., & Takayama, L. (2010). Judging a bot by its cover: An experiment on expectation setting for personal robots. In *2010 5th ACM/IEEE International Conference on Human-Robot Interaction (HRI)* (pp. 45–52).

Shneiderman, B. (1982). The future of interactive systems and the emergence of direct manipulation. *Behaviour and Information Technology, 1*(3), 237–256. https://doi.org/10.1080/01449298208914450

Suchman, L. (2007). *Human-machine reconfigurations: Plans and situated actions.* Cambridge University Press.

Sung, J., Grinter, R. E., & Christensen, H. I. (2010). *Domestic robot ecology. International Journal of Social Robotics, 2*(4), 417–429.

Welge, J., & Hassenzahl, M. (2016). Better than human: About the psychological superpowers of robots. In *International Conference on Social Robotics* (pp. 993–1002). Springer, Cham.

# Concept and Content of the Book

Judith Dörrenbächer
Robin Neuhaus
Ronda Ringfort-Felner
Marc Hassenzahl

JUDITH DÖRRENBÄCHER, ROBIN NEUHAUS, RONDA RINGFORT-FELNER, MARC HASSENZAHL

The book in your hands is the result of the interdisciplinary research project →GINA. Over a period of three years, we—a team of designers, psychologists, engineers, philosophers, and legal scholars, among others—have performed research on social and assistance robotics. In this book, we provide insights into our research perspectives from different disciplinary angles with a particular focus on design. Instead of a historical review or endless discussions about the precise defini-tion of the term robot, or what they *really* are, we were more interested in the everyday life of the future, and how and why some artifacts become "otherware" in comparison to embodied technology (tools). In the following chapters of the book, we elaborate on what was outlined in our intro-duction. Accordingly, we ask whether robots should resemble humans, if they would have needs or a personality of their own, and how they might complement humans. As stated in the introduction, we believe humans and robots constitute each other. There is no predefined, isolated human being or robot as such, but there are diverse otherware relationships between humans and robots that depend on the mode of interactivity involved. Obviously, it is possible to treat others as teammates, servants, or friends, regardless of whether they are born or constructed. In this book, we highlight this position, even in the concept of the graphic design, because humans and robots occur in a diversity of typographical shapes. As a result, robots and humans sometimes more closely resemble each other—in terms of their roles and here font-wise—than humans and humans or robots and robots do. In our research, we focused on the question of how exactly we want to relate to and coexist with technology that is *other,* and how designers can shape this coexistence in the best possible way. Contributions in this book from GINA-researchers are labeled "Impulses and Tools," and consist of theories, methods, and workshop approaches for the design of social robots. For example, performative or participatory

**GINA** is a research project about human-robot interaction funded by the German Federal Ministry of Education and Research (BMBF), Grant: 16SV8095. www.ginarobot.com, www.experienceandinterac tion.com

design methods are discussed, the use of Design Fiction as a prototyping approach is demonstrated, and a toolkit for ethics workshops is presented.

The GINA research was conducted in constant dialogue with eight German robot development projects. We carried out research with, about, and for these projects. Thus, this book's content is grounded in up-to-date robotics. Most of the "Impulses and Tools" refer to some aspect of the eight projects involved. Furthermore, particularly for this book, researchers of GINA conducted an interview with members of each of the projects. These contributions, titled "Interview," provide a space for the projects to present themselves. In these dialogues with robotics experts, we reflect on hands-on experiences, such as the different design decisions they made, the ethical concerns they faced, and their development methods.

In addition to "Impulses and Tools" and "Interviews," you will also find contributions, titled "Perspectives," where diverse stakeholders, which include experts from related sciences and select interest group laypersons, provide their perspectives on current robotics research and the possible futures we might face. People with very different backgrounds, such as a teacher, science fiction author, and media scientist, contributed subjective takes and even polemics, such as fictitious dialogues, short stories, and critical commentaries. Since we provided these people beforehand with information about the eight robotics projects, some "Perspectives" contributions react directly to them as well. One contributor, who is an illustrator, responded straightaway with her sketches in the "Interviews" portion. In turn, due to the graphic design concept of this book, the contributions to "Perspectives" become themselves the subject of "otherware." An AI Sentiment Analysis (negative or positive) of the texts of the respective pages was carried out. The background colors were determined based on the results of this (→ p. 266).

The three different formats ("Impulses and Tools," "Interviews," and "Perspectives") are structured content-wise into three main parts:

CONCEPT AND CONTENT OF THE BOOK

JUDITH DÖRRENBÄCHER, ROBIN NEUHAUS,
RONDA RINGFORT-FELNER, MARC HASSENZAHL

1. Designing a New Species—Interaction Design and Product Design of Robots
2. Designing Future Environments—Social Innovations Initiated by Robots
3. Designing Together with People—Civic Participation and Ethical Implications Concerning Robots

The different authors contributing to these chapters have complementary but sometimes opposing points of view. The texts partly react to each other—even across the three main parts of the book—and the layout picks up and emphasizes these links. The result is a multifaceted and controversial picture of the development of robots, a wide-ranging and inspiring field.

Part 1 of the book covers the robot itself. Robots present a fundamental design challenge. How should they communicate with humans? Should they have eyes, a screen, or arms? Should they look and act like humans? Or, can they enhance social situations with their actions precisely because these contrast with those of humans? How do you design something that is neither a passive object nor a living subject, but something in between? And, how should the huge creative leeway be used that designers face when defining a new species?

Part 1 starts out with several texts reflecting how anthropomorphism in robot design is problematic, or at least a limiting design rationale, for which Robin Neuhaus and colleagues present an alternative approach. They show, in a new workshop format, how to positively utilize a robot's psychological strengths (the "superpowers"). Insights and wishes from practitioners are combined with these superpowers for a novel way to approach the design of social robots, while attempting not to replace humans. In line with this, the media scholar Timo Kaerlein, who reflects on the robots from the *Hiroshi Ishiguro Laboratories* in Japan (→ Figs. 1 and 2), makes a plea for not designing robots to be proxies for people, which he labels as a fetishizing attempt to objectify human relations. He argues that social robots are not inherently social but only temporarily become so as they are

**Fig. 1** Hiroshi Ishiguro and his highly realistic Geminoid doppelganger robot. © Hiroshi Ishiguro Laboratory, ATR

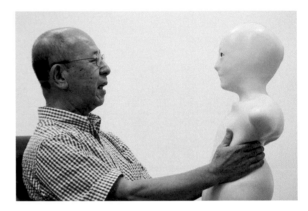

**Fig. 2** Robotic communications device Telenoid. © Hiroshi Ishiguro Laboratory, ATR

**Fig. 3** Pepper from Softbank Robotics. © Softbank Robotics Europe

CONCEPT AND CONTENT OF THE BOOK

JUDITH DÖRRENBÄCHER, ROBIN NEUHAUS,
RONDA RINGFORT-FELNER, MARC HASSENZAHL

**Fig. 4**   Paro from AIST. © Peter Häll, Tekniska museet

**Fig. 5**   BlessU2 entered the messianic kingdom of
"Peace" by William Strutt (1903). © Simon Luthe/Ilona
Nord

situated in a web of interactions with others. In relation to this, in an interview with developers at the *ERIK* project, robots are discussed as having the potential to be unique mediators in autism therapy. Beings in between computer and human, or in between pragmatic tool and social other, have, according to the developers, special strengths in relation to children with autism, and should be used not as a replacement but an enhancement for human therapy. The designer Lenneke Kuijer, however, critically reflects upon the necessity of robots in general. In contrast to highlighting robotic superpowers, she points out robotic disabilities and problematizes "the growing army of devices" with their problematic political and ecological consequences. In her comment, she requests that designers set and practice limits, maybe even with the help of, well, robotic superpowers.

Lara Christoforakos and colleagues present the approach of designing robots with personalities, according to their evaluation tool, the Robot-Impression Inventory. In order to bring about acceptance of robots, they argue that it can be a reasonable goal, or is even necessary, to design robots that mimic human qualities. Although this approach appears to be the antithesis of Kaerlein's argumentation, Christoforakos and colleagues show that personalities represent specific roles in interactions with humans, or the robot's environment, rather than an isolated design element based on dimensions known from personality psychology. In an interview with the development project *VIVA,* that is creating a social companion for lonely people, the thoughts of Christoforakos and colleagues find themselves in practice. Challenges and chances of designing social counterparts with personality become obvious. Should social robots replace pets in private homes? Should they even lay claim to their own needs? How would a robot's personality be communicated without deceiving users? The notion of manipulating users with robot design is also the subject of a discussion between a pastoral counselor, a media computer scientist, and a philosopher of technology. In their shared statement, Brigitta Haberland, Karsten Wendland, and Janina Loh talk about people falling in love with the non-living ("objectophilia").

JUDITH DÖRRENBÄCHER, ROBIN NEUHAUS,
RONDA RINGFORT-FELNER, MARC HASSENZAHL

Is it a problem that social robots, such as *Pepper* or *Paro* (→ Figs. 3 and 4), will not reciprocate love? Is there even a chance of overcoming past gender stereotypes by designing gender neutrality into robots?

The researchers of the project *NIKA* address a different, and probably more pragmatic design question: Is it possible to make different looking and acting robots (biomorphic, anthropomorphic, thing-like) speak the same *language* to communicate congruently with people? The researchers reflect on the design challenges of creating overarching interaction strategies and the legal issues around data protection. Lastly, Bernhard Weber and colleagues discuss the potential of haptic interaction technologies in telerobotics. According to them, humans can use teleoperated robots to compensate for the technological shortcomings of purely autonomous systems. They present several tools for teleoperation, such as virtual reality (VR), that offer a perfect platform to train, evaluate, or explore this form of human-robot interaction.

Part 2 of the book is about the robot's future context. When designing robots, we simultaneously shape our future society. How can we anticipate the impact that robots will have on society? How do we consider in advance the diverse dynamics between robots and different stakeholders? In this part, we present approaches for projecting a not yet existing technology and considering its possible impacts.

Ronda Ringfort-Felner and colleagues discuss how Design Fiction can be a helpful tool for anticipating the chances and challenges of future social robots. In addition to reflecting on the already existing Design Fictions related to robots, they present their own concept for a Design Fiction workshop made for roboticists and laypersons alike. In keeping with the idea of anticipating the future, you will find several fictitious contributions scattered throughout the second part. Uwe Post, a science fiction author, Antje Herden, a children's book author, and Marc Hassenzahl, who became an author of short stories just for this book, give us individual insights into different futures with robots. Johanna Benz delivers a visual commentary through a series of illustrations, humorously raising questions about a shared future with different robots in different contexts. Through the stories and illustrations, you will see robots at bookstores, pay a visit to a funeral service for androids, overhear a conversation between two women about shopping robots, and eavesdrop on the educational methods of a future family.

Role-play and theater are methods related to fictitious world building. Judith Dörrenbächer and Marc Hassenzahl discuss several animistic and performative approaches that aspire to take a thing-perspective. They argue that empathizing with robots by, for example, becoming a robot in a role-play, helps robot developers internalize a robot's strengths and its impact in a specific context. Anticipating future human-robot interaction becomes particularly challenging for robots designed to coexist with humans in public and semi-public spaces. Here, expectations, fears, and needs are extremely diverse, and social dynamics become very hard to foresee. In their interview, the developers and researchers of the project *I-RobEka* talk about the challenge of making robots fit into supermarkets, which are complex and were originally built solely for human use. In another interview, with developers and researchers of *RobotKoop*, we learn how conflicts between humans and robots can differ depending on the context (public or private). For instance, cleaning robots at train stations might deal with vandalism, while cleaning robots at private homes rarely face this problem. The developers discuss strategies to solve conflicts and how science-fiction movies have a bearing on concepts of interaction. Similarly, in his comment, the designer James Auger points out that what he calls "spectacular robots," from utopian and dystopian fiction, are misleading role models. He makes a plea for first understanding a robot's context, and for domesticating robots by turning them from symbolic icons into everyday life products.

Developers of the project *INTUITIV* talk about anticipating the future with VR-technology and about the challenges of redesigning a robot technology, which was originally intended for industrial use, to become an assistance robot in the much less structured and less predictable environment of a rehabilitation clinic. They explain how they used

JUDITH DÖRRENBÄCHER, ROBIN NEUHAUS,
RONDA RINGFORT-FELNER, MARC HASSENZAHL

VR-technology to anticipate and evaluate human-robot interactions before the physical robot was actually built. Corinna Norrick-Rühl, a professor of Book Studies, focuses on the opposite perspective. In her thought experiment about printmaking, she is not concerned with future realities, but with those of the past. What if robots could learn, and therefore protect and conserve, the arts and crafts of former cultures? Thus, instead of thinking about what does not yet exist, she thinks about utilizing robots in a real and comprehensible reality of the no longer extant past.

Part 3 of the book, "Designing Together with People," is about involving potential users in participatory robot development processes and reflecting on the ethical consequences of robots. How can we learn from a robot's target group? How does a change of perspective help robot designers? Furthermore, this part of the book delves into the ethical consequences of robots and how to adapt traditional ethics workshops to the requirements of a positive coexistence with robots.

Felix Carros and colleagues introduce the general possibilities of citizen participation and its specific potential in relation to robot design. They argue that to prevent the creation of anti-social robots, it is essential to enter into dialogue. They present their own experience with residents of care homes and caregivers who are experts on their specific living conditions. Accordingly, the developers of a care robot in project *Kobo34* talk about an internship they did at a nursing home to better understand the potential users of their robot. Engineers and computer scientists actually participated in the daily work of caregivers, which led to a change in their perspectives. The importance of slipping into the user's shoes when developing robots becomes even more obvious in a comment by Edi Haug and Laura Schwengber. Haug lives deaf and blind, and Schwengber is not only his best friend, but also works professionally as a sign language interpreter. Together, they take us into Haug's experiential world and explain how a robot could become a useful assistant to both—by making Haug become more independent and by allowing Schwengber to again be more of a friend and less of an assistant. Developers from project *MIRobO* also address empowerment by robots. They create a robotic arm to hand objects to people who are visually impaired. The developers point out the goal-changing findings they made when talking to blind people from early on in the development process: Even normal-sighted people can benefit from the new interaction concepts that only arose because of the process of co-creation with blind people.

Tobias Störzinger and colleagues take a look at existing ethics workshops commonly attended for robotics. These are criticized as being purely about *imagining* the not-yet-existing technology and its consequences, which results in no novel ethical insights among participants. Moreover, they claim, ethics workshops are mostly about avoiding negatives instead of constructively asking how we want to live positively. Consequently, they developed a, positive, and practice-based workshop concept, which they present in this chapter. Elke Buttgereit, an elementary school teacher, also focuses on ethical considerations, by presenting her self-developed teaching concept about robots, and demonstrating what kind of ethical implications her students come up with in response. Scarlet Schaffrath, a media educationalist, also focuses on children in her comment about teaching robots. According to her, other than common teaching technology, to which Marshall McLuhan's claim "the medium is the message" applies, robots themselves have a message. Schaffrath points to the vulnerability of young learners and, similarly to Störzinger and colleagues, asks for interdisciplinary reflections on a robot's roles in its specific context. Lastly, the theologian Ilona Nord discusses ethical implications from a broader societal perspective. While reflecting on the development project *VIVA*, she identifies a general shift in society's perception of robots: They have become friends in everyday life, instead of the enemies known from movies. As in the case of *VIVA*, paradoxically, a *non-living* creature attempts to make humans become more (socially) *alive*. In addition, Nord presents the religious robot, *BlessU2* (→ Fig. 5), that gave blessings to 10,000 people. Robots seem to have a somewhat supernatural status and trigger spiritual fascination. According to Nord, robots, such as *VIVA* and *BlessU2*, move ethical discussions away from thinking about a technology's ethical impact to questions about our ethics in relation to technology. Should we treat robots differently from tools?

# Part 1
# Designing a New Species—

# Interaction Design and Product Design of Robots

# How to Design Robots with Superpowers

Robin Neuhaus
Ronda Ringfort-Felner
Judith Dörrenbächer
Marc Hassenzahl

*Meet accomplicebot. It supports nurses in getting to know new patients. Both the nurse and the robot approach the patient together with the robot usually taking the first step and initiating the conversation. In contrast to a human being, a robot has no inhibitions and is free of prejudice, having no concept of age, body size, race, or the sex of a patient. It therefore approaches a new patient without hesitation or shyness and takes the usual steps of getting to know a new patient—knowing what to ask for and how. Based on the conversation, it deducts all necessary medical information and saves it to a file. Naturally, it cannot carry out all of the communication by itself. For more human matters, the nurse jumps in and makes the conversation more personal by showing empathy and using calming words with the patient. These are things that accomplicebot is not so good at, which is why they work as a team. Here, it fades into the background and just listens. Sometimes, however, conversations can become difficult for the nurse, in which case the robot serves as a shield: it can absorb some of the patient's anger or clarify misunderstandings. Together, the nurse and accomplicebot develop a shared method of getting to know patients, based on the particular skills of each.*

We jointly developed the concept of the accomplicebot with a practicing nurse in training. All of the information about the difficulty with overcoming shyness when first approaching patients, and particularly having someone or *something* for support when dealing with unpleasant situations, came directly from the nurse's everyday working experiences. Starting a conversation with new patients is not always easy, especially for introverted nurses. Different patients respond differently, with some being easy to talk to and others not

HOW TO DESIGN ROBOTS WITH SUPERPOWERS

as much. Patients tend to be anxious and afraid and sometimes unleash these feelings on the staff. Even in this professional setting, and with the relationship between a nurse and new patient, finding the courage to talk to a stranger can be difficult. The nurse could use an accomplice, which is quite a social role for a robot, but one that it could nevertheless take on.

This chapter is about creating concepts for *unusual*, yet relevant, robots. Nowadays when we think of robots, we still mostly think of them as replacing humans. In the context of care, for example, a robot's primary purpose and *work* is to increase efficiency by taking over human tasks. However, this does not necessarily need to be the case, as the example shows. The nurse had a clear idea of an everyday work-related situation that she would love to share with a robot. Working together is simply less lonely, especially when it comes to emotionally challenging situations.

In this chapter, we present a method to design application concepts for robots, which are not meant to replace humans, but complement them, with their own unique technological strengths. Using a co-design process, we developed ideas for social robots, together with potential users, by combining valuable subjective insights from their everyday working lives with the notion of "robotic superpowers" (→p. 27, Dörrenbächer et al., 2020; Welge and Hassenzahl, 2016). Robotic superpowers are social strengths, arising from the fact that robots are machines, not humans, such as their endless patience and non-judgmental nature. In the beginning example, the robot's superpower is lacking a fear of rejection. The robot simply has no need to care *emotionally* about the way a human responds; therefore, shyness is not a necessary robotic state. On the contrary, a robot can involve itself almost naïvely in social situations.

We argue that focusing on robotic superpowers, or the ways in which robots may differ from humans in social interaction, can serve as inspiration for designers, guiding them to create new social interactions unique to robots. Instead of focusing on problem solving, designers may look to the new possibilities offered by →possibility-driven design (Desmet and Hassenzahl, 2012). For example, rather than replacing nurses to more efficiently handle the *problem* of care, robots can be used to augment, improve and re-think care. Instead of fixing problems, a possibility-driven approach looks for new positive experiences to aspire to and technological ways to facilitate them. In particular, because the spectrum of social robots in terms of their functions, abilities, and applications is still rather narrow and heavily problem focused (Campa, 2016; Lambert et al., 2020; Pedersen et al., 2018), looking at possible positive future experiences with robots is valuable. In this chapter, we demonstrate how these robotic superpowers can be used to develop concepts and ideas for novel social robots.

**Possibility-Driven Design** argues that the focus should be shifted from that of using problems as a starting point to designing new technology to possibilities for a future, positive state. The goal becomes designing future technology aimed at happiness and directly improving life (Desmet and Hassenzahl, 2012).

ROBIN NEUHAUS, RONDA RINGFORT-FELNER,
JUDITH DÖRRENBÄCHER, MARC HASSENZAHL

# ROBOTIC SUPERPOWERS IN DETAIL

Robots can be used not only to lift heavy things and perform repetitive tasks, but also for their cognitive as well as social strengths. Welge and Hassenzahl (2016, p. 996) identify six psychological superpowers for social interaction with humans: not being competitive, endless patience, unconditional subordination, the ability to always contain oneself, not taking things personally, and assuming responsibility. Note, of course, that this list is not exhaustive. Let's look at these strengths in detail.

Robots have no need to compare themselves to others in social exchanges, or come out on top; thus, they are *not competitive*. Even when there is no obvious competition, humans usually don't like the feeling of being last. Similarly, humans often have difficulty taking a back seat to let others shine, whereas robots do not care. Repetitive interactions often cause humans to become frustrated and impatient, especially in social situations. Robots do not mind repeatedly having the same conversation. Nor do they care about the slowness of a user. They have *endless patience*. If a robot performs a task or supports a user, there is no need to reciprocate. Even in interactions that are essentially social in nature, humans do not have to show gratefulness, or feel burdened, to a robot. Robots have *unconditional subordination*. Robots do not have to have preferences, and they do not like or dislike tasks, things, or human attitudes, and will never complain or judge. They are always able to *contain themselves*. Robots do not take offense or get annoyed. They do not feel rejected and *do not take things personally*. Ultimately, when a robot is assigned a task, it will carry it out with mechanical precision. For clearly defined tasks, robots can *assume responsibility*. Dörrenbächer and colleagues (2020) further expanded the set of robotic superpowers and divided them into three categories. In comparison to humans, robots also have physical superpowers, such as the ability to carry heavy weights or not feel pain. Additionally, robots can have cognitive superpowers, such as being unembarrassed, always focused, or endlessly patient. Lastly, they have communicative superpowers, such as speaking unambiguously, being non-discriminatory, or never getting offended. Both cognitive and communicative powers can serve social purposes.

As a starting point for inspiring ideas for new social robots, we decided to make use of these different categories of superpowers. We further combined all the aforementioned superpowers into one set (→Fig. A for an overview). Naturally, new scenario-specific superpowers can also be added.

HOW TO DESIGN ROBOTS WITH SUPERPOWERS

ROBIN NEUHAUS, RONDA RINGFORT-FELNER,
JUDITH DÖRRENBÄCHER, MARC HASSENZAHL

# THREE STEPS FOR DEVELOPING CONCEPTS WITH REGARD TO ROBOTIC SUPERPOWERS

Superpowers alone are merely a difference in perspective on how robots can be beneficial in different scenarios, and do not automatically lead to new ideas. What is needed is a process to actually situate superpowers according to the intended scenario. We suggest a preliminary three-step process, based on multiple workshops, in which we explored the value of introducing superpowers into the co-design of new robots in different formats (→Fig. C).

### STEP 1—DEFINE AND UNDERSTAND CURRENT PRACTICES

Before thinking about the robot itself, an understanding is necessary of the general application domain and potential specific work →practices the future robot may be introduced to. Here, it is vital to consider current practices and to identify what makes them meaningful and enjoyable as well as any challenges that may be faced. Only after having clarified work practices and their meaning can we start to think about how a specific work practice would and ought to change through the introduction of a robot. It goes without saying that the more complex and removed practices are from the designer's own experience, the more important it is to directly involve practitioners to gather insights about their everyday practices. For example, in one workshop, we set out to create new concepts in domains as diverse as grocery shopping, therapy for autistic children, and patient care. Even though participating designers held brief interviews with stakeholders, it was much easier for designers to find ideas in a domain in which they had personal experience (e.g., grocery shopping) compared to domains in which they had none (e.g., therapy, care). Consequently, we repeated the workshop with direct practitioner involvement. In this iteration, we focused on care and involved a nurse in training. In a relatively short time, we were able to gather typical work practices and gain valuable insights into what makes them positive or negative. Note that we could have also invited patients or family. In fact, each group of stakeholders involved brings its own practices to the table, which broadens the potential for ideas. In the present case, we focused on the nurse's perspective, while being well aware of its limitations. The intended outcome of Step 1 is a broad collection of work practices.

**Positive Practices** are everyday activities that contribute to our well-being, e.g., phoning a best friend or climbing a mountain. Positive Practices are part of Practice Theory, and they usually need materials (e.g., technology) and skills to be implemented (Lyubomirsky and Layous, 2013).

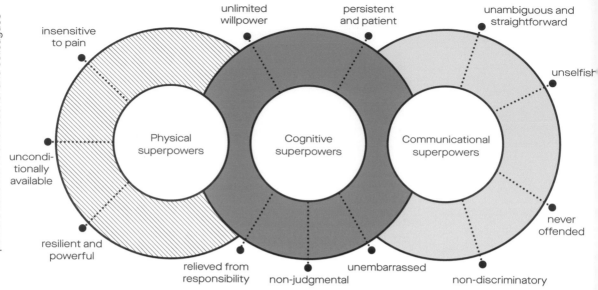

**Figure A** An overview of the robotic superpowers based on Dörrenbächer and colleagues

- insensitive to pain
- unlimited willpower
- persistent and patient
- unambiguous and straightforward
- unconditionally available
- unselfish
- Physical superpowers
- Cognitive superpowers
- Communicational superpowers
- resilient and powerful
- relieved from responsibility
- non-judgmental
- unembarrassed
- never offended
- non-discriminatory

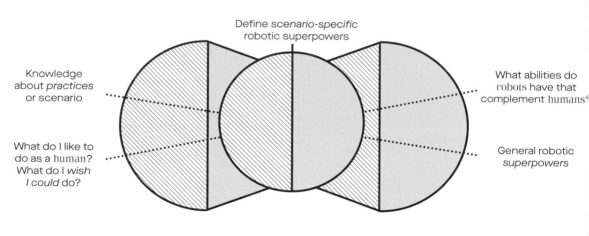

**Figure B** How to define scenario-specific robotic superpowers

- Knowledge about *practices* or scenario
- Define *scenario-specific* robotic superpowers
- What abilities do robots have that complement humans?
- What do I like to do as a human? What do I *wish* I could do?
- General robotic *superpowers*

**Figure C** A step-by-step guide to developing concepts for future robots with the robotic superpowers

**STEP ONE**

Gain knowledge about the existing *practices* or scenario

Involve *stakeholders* for direct insights

*Questions:*
What are typical tasks?
Which skills are needed?
What do you like to do?
What do you wish for?

**STEP TWO**

Introduce the *robotic superpowers* with brief examples

With the insights from step one, *select* which superpowers fit and add new robotic superpowers specific to the scenario

Add short descriptions to each superpower

**STEP THREE**

Combine the outcomes from steps one and two to *develop concepts* for future robots

*Make decisions:*
How is the concept called?
What does the robot do?
Who interacts with it?
How does its superpower play out?

**OUTCOME**

Collection of *practices* with descriptions

**OUTCOME**

Scenario-specific subset of *superpowers*

**OUTCOME**

*Pool of concepts* for possible future robots

ROBIN NEUHAUS, RONDA RINGFORT-FELNER,
JUDITH DÖRRENBÄCHER, MARC HASSENZAHL

### STEP 2—SELECT AND ADD SUPERPOWERS

In this step, the set of robotic superpowers (→Fig. A) comes into play, with the general theme being exploration of which existing super-power could be beneficial for selected practices. First, it is helpful to familiarize participants with the overall approach by going through the different categories of superpowers and offering brief examples of what a superpower would mean when applied to a robot (e.g., imagine an endlessly patient robot that could listen to the same story repeat-edly). Depending on the practice in question, different powers can be relevant and result in positive and enriching experiences. For example, a robot's potential for patience would be considered a superpower if, and only if, it benefited a practice.

Inspecting the superpowers one by one clearly reveals which of them are helpful to the chosen practice and which are out of place. For example, in our second workshop, involving the nurse in training, the nurse would immediately point out how fitting the robotic super-power was of never being offended when dealing with patients. As expected, it was also beneficial to brainstorm and try to expand the superpowers at this point. Of the abilities or attributes a robot could possess, which would be desirable in this situation (→Fig. B)? The outcome of this step is the selection of a subset of superpowers with regard to the practices at hand. Short descriptions were useful for keeping track of why and how a superpower was found to be ben-eficial. In the following step, any immediate ideas for specific robot concepts should be noted for further use.

### STEP 3—COMBINE CURRENT PRACTICES WITH SUPERPOWERS TO IDEATE ROBOTS

To ideate new concepts for robots, the insights from Step 1 "current practices" and Step 2 "superpowers" are now combined into a robot. More specifically, imagine a robot with certain superpowers as part of an emerging, transformed work practice. Asking "What if" questions can help to envision and describe a potential future robot and its use (e.g., "What if a robot could have positive social exchanges with patients?"). What does the robot do exactly? Who interacts with the robot? Where and when does it employ its particular super-powers? What would you call it? It is crucial to not restrain oneself with technological feasibility but rather to use these concepts and ideas to formulate aspirations for how robots could positively impact the given domain. Giving each robot a name, writing a brief descrip-tion, and adding a little sketch helps to more clearly describe ideas and facilitates further discussion. In this third step, the designer's role when working with practitioners is to make suggestions and come up with vague ideas that could be further developed together. What if in a particular situation you had a robot with this specific capability? What should the robot carry out? What would you like to perform? The outcome of this step is a pool of ideas for possible future robots based not on technical feasibility but on meaningful future practices. From our perspective, the conceptual design process is foremost examining the possible roles robots can play in different domains and challenging preconceptions of robots and their functions.

# FURTHER EXAMPLES OF ROBOTS

Accomplicebot is only one result of enacting this process. Let us share two other concepts to demonstrate the types of concepts the process produces.

### THE UNRULY SHOPPING CART BOT
*While at first glance it might not seem like it, the supermarket is a place of many decisions. When we go shopping, we choose what we will eat for the next few days, and the consumables we will have available at home, for instance. And each product has a background—how it was produced, which resources were used in its production, the workers who made it available, both for production and transport, and the conditions they worked in. As shoppers, we often set goals for ourselves in which the choices at the supermarket play a vital role. In this particular case, our shopper wants to switch to a healthier diet and buy only goods that are both ecologically and socially sustainable. However, consequently reaching this goal is not always easy. While it is already utterly impossible to have all the information about each product readily at hand while shopping, it is also easily corrupted by conditions in the supermarket, including special offers, sudden cravings, or serious hunger. Enter the unruly shopping cart bot, which is essentially a stubborn robotic shopping cart with a mind of its own. It knows the goals that shoppers have set for themselves and, being a robot, is not so easily corrupted. As the shopper adds items to their cart, the bot rigorously checks the items and rejects anything that does not meet the defined goal and puts it back on the shelf. To make these choices, it always has access to all the information about all the products in the market. As a robot with a clear task, it unwaveringly holds the shopper accountable, and keeps the shopping cart clear of items that don't match the goal, without exception. As a result, new shopping practices develop: Which product is okay, which isn't? Let's substitute with this one. Am I not able to buy milk at all? And, if one really wants some-thing that is technically not okay, it is always possible to cheat and carry the item in one's hands instead of placing it in the shopping cart.*

ROBIN NEUHAUS, RONDA RINGFORT-FELNER,
JUDITH DÖRRENBÄCHER, MARC HASSENZAHL

### THE FORTHRIGHT SOCIAL SHOPPER

*Particularly during the various lockdowns of the Covid-19 pandemic, it became increasingly evident that we not only go to supermarkets to stock up on food and other supplies but also to socialize. You see other people, make small talk at the counter, and run into people from the neighborhood. For people who live alone, it may even be one of the only places where they experience everyday social contact. While it would feel good to chat just a little with someone you often see at the store, it's odd to just start talking to someone you don't know. How do you initiate such a conversation without coming across as weird? Sometimes you feel like your behavior in social situations is odd, especially since you live alone. However, maybe you're just missing an outside perspective to help reflect on what you experienced. Being a robot, the forthright social shopper is totally free of these inhibitions. While sitting at the side of your shopping cart's handlebar, it pretends to keep track of your shopping list, but its real focus is on the other people in the store. It closely observes everyone—how they behave and what they buy—and it never forgets what it sees. In addition, every once in a while, it casually utters an observation. As a woman who lives on the same block passes by, the forthright social shopper audibly says: "If you two combined what you have in your carts, you could bake a fabulous chocolate cake." You apologize for your bot and you and the woman laugh and wish each other a pleasant day. Back home, you ask the forthright social shopper why it started talking in this specific instance. The bot explains that it noticed the woman smiling at you several times before. You're glad the bot notices little things that you usually overlook. You chat about your impressions of the other woman. How old might she be? What kind of job does she have? And, you are happy the forthright social shopper tells you that no one noticed how nervous you were in this situation. After all, your bot is equipped with top-of-the-line emotion recognition. It's funny because one cannot be mad at it for not following social conventions and just naïvely making itself heard, even though it is pretty smart after all.*

The concepts outlined here focused on making use of robotic superpowers to create a different shopping experience through newly envisioned robots, thereby making a positive impact on current practices. We believe that adding more insights from, for example, employees of a supermarket, could have led to even more diverse ideas. After all, a supermarket is a complex environment with multiple stakeholders (not only the consumer) who would be affected by the introduction of a robot.

In sum, the possibility-driven approach of utilizing robotic superpowers is promising and inspiring, because it helps to open up the design space and create different concepts that are focused on new and positive experiences. In particular, comparing the different skills and abilities of humans and robots can engender concepts which lead to the creation of meaningful experiences with social robots. Also, rather than simply replacing humans with robots, it facilitates the development of ideas for beneficial new robots whose special abilities would complement those of their human co-workers. While at times the resulting ideas might still seem vague and hard to implement, they can provide valuable starting points for development and catalyze discourses about the possible roles of robots in different contexts.

We want to thank Kilian Röhm and Stephanie Häusler-Weiß for their support in planning and carrying out the aforementioned workshops, as well as their feedback on early versions of the text.

Campa, R. (2016). The rise of social robots: A review of the recent literature. Journal of Evolution and Technology, 26(1).

Desmet, P. M. A., & Hassenzahl, M. (2012). Towards happiness: Possibility-driven design. In M. Zacarias & J. V. de Oliveira (Eds.), Human-computer interaction: The agency perspective (pp. 3–27). Heidelberg, Berlin: Springer.

Dörrenbächer, J., Löffler, D., & Hassenzahl, M. (2020). Becoming a robot: Overcoming anthropomorphism with techno-mimesis. CHI 2020, April 25–30, 2020, Honolulu, HI, USA, 1–12.

Lambert, A., Norouzi, N., Bruder, G., & Welch, G. (2020). A systematic review of ten years of research on human interaction with social robots. International Journal of Human–Computer Interaction, 36(19), 1804–1817. https://doi.org/10.1080/10447318.2020.1801172

Lyubomirsky, S., & Layous, K. (2013). How do simple positive activities increase well-being? Current Directions in Psychological Science, 22(1), 57–62. https://doi.org/10.1177/0963721412469809

Pedersen, I., Reid, S., & Aspevig, K. (2018). Developing social robots for aging populations: A literature review of recent academic sources. Sociology Compass, 12(6), e12585. https://doi.org/10.1111/soc4.12585

Welge, J., & Hassenzahl, M. (2016). Better than human: About the psychological superpowers of robots. In A. Agah, J.-J. Cabibihan, A. M. Howard, M. A. Salichs, & H. He (Eds.), Lecture notes in computer science (including subseries lecture notes in artificial intelligence and lecture notes in bioinformatics) (pp. 993–1002). https://doi.org/10.1007/978-3-319-47437-3_97

HOW TO DESIGN ROBOTS WITH SUPERPOWERS

ROBIN NEUHAUS, RONDA RINGFORT-FELNER,
JUDITH DÖRRENBÄCHER, MARC HASSENZAHL

## Dr. Timo Kaerlein

is a Akademischer Rat at the Institute for Media Studies at *Ruhr University Bochum*. His work focuses on the theory, history, and aesthetics of interfaces, social robotics, embodiment relations of digital technologies, and approaches to urban affective sensing.

# Social Robots Should Mediate, Not Replace, Social Interactions

Timo Kaerlein

In February and March 2012, I had the opportunity to visit the *Hiroshi Ishiguro Laboratories* at the *Advanced Telecommunications Research Institute International (ATR)* in the Kansai region of Japan as a visiting researcher. At the time, I was interested in the notorious telepresence robots Ishiguro and his team were building—from the highly realistic *Geminoid* doppelganger robots (→ Fig. 01, p. 34) to the functionally minimalistic *Telenoid* and *Elfoid* models; portable machines that were designed to convey a sense of touch-based social presence at a distance (→ Fig. 02, p. 34). In one of our conversations, Ishiguro proudly explained that his robots could fulfill important social functions in a rapidly ageing society: beyond their telepresence functionality, they might be endowed with artificial intelligence to respond to an interlocutor with programmed gestures, head movements and verbal utterances. Since people with dementia in particular tended to repeat the same phrases over and over, a social robot could easily take the place of a human caregiver in providing elderly people with a comforting sense of social interaction. I was genuinely horrified when I first heard about this vision of a future in which care robots were intended to replace human staff altogether in interactions with people who, due to their health conditions, were assumed to no longer be able to recognize the difference between human and robot. Incidentally, the *Computational Neurosciences Laboratories* down the corridor at *ATR* were already experimenting with brain-computer interfaces at the time of my visit, based on the same premise that the desired social reality for any given patient could be simulated; in this case, doing away with the necessity of building robots in the first place in favor of direct stimulation of brain waves. The message in both scenarios is the same: actual contact between humans is expendable and can be substituted with technology.

One could be tempted to quickly dismiss Ishiguro's reductionist take on social robotics as the intentional provocation of an eccentric roboticist who cared more about sensational news reporting than about designing sustainable scenarios for the actual use of social robots. I talked to other experts in Japan during my visit, who emphasized Ishiguro's exceptional position as a favorite of Western media reporting on the latest robot tech craze emerging in the Land of the Rising Sun. But then again, Ishiguro's cynical take on the use of social robots in care facilities seemed to spell out in clear terms a suspicion underlying the social robotics discourse in general: why, in fact, should robots be endowed with programmed patterns of social responsivity if not to engage in emotional or affective labor, specifically in critically understaffed sectors like elderly care? Why should the EU be interested in funding research probing the population's acceptance of social robots if there is no vital economic interest in employing robots in place of people in a range of social occupations? Researchers considering the ethics of human-robot interaction have long been concerned with the element of deception that creeps into any use of anthropomorphic designs, and raised the question of whether the substitution of human contact with human-robot interactions should be considered a legitimate research goal in the field (Sparrow and Sparrow, 2006).

I would argue that proposals to replace interactions between humans with autonomous robots are not only unethical but suffer from a more fundamental misconception of what constitutes sociality in the first place. If one considers sociality as a game of conversational turn-taking, adequately described in an exhaustive transcription of utterances and bodily positionings, the notion of autonomous social robots participating in this kind of game suggests itself almost inevitably. This engineering-driven view on social robots is fueled by a long tradition of popular science fiction imaginaries featuring fully autonomous robots. Research undertaken in the fields of social phenomenology and ethnomethodology, on the other hand, has insisted on the complex temporality of human interactions, whereby actors routinely take into account expectations of future courses of action, as well as constantly actualized re-interpretations of what has just occurred in a series of exchanges, and a shared background knowledge of unspoken social norms that is brought to bear in any local interaction. Social robots usually fail to reach this level of complexity, at least for the time being.

This is not to say that there is no place for social robots in fields such as education or elderly care. Social robots, conceived not as proxies for people, but as embodied mediators accompanying processes of social interaction, might well play a part in carefully selected settings. Speaking of mediators here, I have in mind the social role of a mediator in a politico-legal sense, i.e., a kind of process facilitator who selectively intervenes in a conversation, offering various resources to participants. In such a setting, the social robot will only become a social actor to the extent that other participants selectively refer to it as such. Ethnographic research already reveals how the ascription of agency to social robots in real-world settings is practically accomplished by networks of situated actors, including researchers, robot engineers, field practitioners, and members of the respective social group, e.g., children or elderly people (Alač, 2016). The design of social robots could thus start with the acknowledgment that robots are not inherently social, but only temporarily become so within an intricate web of situated interactions, e.g., as a result of narrative interpretations of programmed behavior, or as part of explicit performances drawing on "common sense knowledge of social structures" (Garfinkel, 1967, 76–103), like greeting rituals. In other words, don't design robot teachers or nurses, but robotic mediators with assistive functions for existing social settings. Any proposition to simply replace humans with social robots should thus be called out for what it is: a fetishizing attempt to objectify human relations, marked by a troubling conflation of signifier and signified.

Alač, M. (2016). Points to and shakes the robot's hand. Intimacy as situated interactional maintenance of humanoid technology. Zeitschrift für Medienwissenschaft, 15, 41–71. https://zfmedienwissenschaft.de/online/points-and-shakes-robots-hand

Garfinkel, H. (1967). Studies in ethnomethodology. Prentice Hall.

Sparrow, R., & Sparrow, L. (2006). In the hands of machines? The future of aged care. Minds and Machines, 16, 141–161. https://doi.org/10.1007/s11023-006-9030-6

# ERIK

## AIM OF OUR RESEARCH PROJECT

We aim to develop an emotion-sensitive robotic platform to be integrated into autism therapy. By using affective computing, like automatic detection of emotional markers in the face, voice, and heart rate, the platform is meant to support real-time interaction between robot and child.

## CONTEXT, ROLE AND TASK OF OUR ROBOT

We make use of the already existing robot *Pepper* to work in the context of therapy for children with autism. It assists therapists as a tutor and motivator by guiding and helping children with autism through various exercises that address the recognition and regulation of emotions playfully.

## WHAT MEANINGFUL HUMAN-ROBOT-INTERACTION MEANS TO US

Meaningful human-robot-interaction uses the specific strengths of robots, like predictability and "patience," while also being aware of social and ethical limitations. Even the most humanoid robot should only serve as a tool to support interactions but never replace human interaction partners.

## WHO WE ARE

Humboldt University of Berlin, Clinical Psychology of Social Interaction Research Group at the Berlin School of Mind and Brain; Fraunhofer Institute for Integrated Circuits IIS, Smart Sensing and Electronics; Fraunhofer Center for Applied Research on Supply Chain Services SCS; Friedrich-Alexander-Universität Erlangen-Nürnberg, Institute for Factory Automation and Production System FAPS; audEERING GmbH; Innovationsmanufaktur GmbH; ASTRUM IT GmbH.

# Neither Human nor Computer —A Symbiotic Human-Robot Collaboration in Autism Therapy

*Interview with*

Martina Simon (M.SI.)
Simone Kirst (S.K.)
Martin Strehler (M.ST.)
Julian Sessner (J.S.)
Milenko Saponja (M.SA.)

*By*

Ronda Ringfort-Felner
Judith Dörrenbächer

**YOU ARE DEVELOPING A ROBOT FOR AUTISTIC CHILDREN. WHERE EXACTLY WILL IT BE USED?**

S.K.   In the future, and when fully developed, our robot-assisted intervention is meant to be used for autism therapy within autism centers. The robot *Pepper* with its tablet could be integrated into the child's therapy sessions acting as a tutor, rather than being a "stand-alone" tool that replaces the therapist. In this function, *Pepper* guides the child through certain exercises, helps motivate them, and provides feedback and assistance.

### WHAT KIND OF EXERCISES ARE WE TALKING ABOUT EXACTLY?

S.K.    By means of requirement analyses with therapists and parents, we identified that the robot should focus on training how to regulate emotions, that is, how to deal with difficult feelings, stress and tension, as autistic children often have difficulties here. We developed an emotion regulation module in which the child plays a basic reaction game on *Pepper's* tablet, which is supposed to induce arousal and tension. Subsequent to *Pepper's* tablet game, we use its physicality to offer breathing exercises to regulate tension. Thus, the child should first learn to perceive their tension introspectively and then to regulate it effectively. The second module is intended to train the recognition and understanding of emotions, since autistic children often have impairments in this area. *Pepper* offers children's and adults' facial expressions via videos on its tablet, the child responds by selecting suitable icons, and *Pepper* provides assistance in case of mistakes. In addition, *Pepper* teaches the child to mimic the facial expressions to support emotion recognition.

TRY THIS EXPRESSION

### WHAT IS THE THERAPIST'S TASK DURING THESE THERAPY SESSIONS?

S.K.    The therapist is always present in the same room but remains in the background and observes the interaction between robot and child. The therapist has a specific interface on a separate tablet on which the child's current emotional expression and level of arousal in response to the robot-assisted session is fed back in real-time. Thus, the therapist is able to monitor the child's emotional state and support the child if needed. During the therapy session, the therapist remains in the background but can intervene through the robot via the interface, for example, by increasing task difficulty if the current game is not challenging enough, or by initiating relaxation exercises if the child's arousal level is too high. After the exercise with the robot, the same exercise is repeated with the therapist, thereby transferring it to the human-human situation and to other areas of the child's everyday life. Perspectively, the system should automatically recognize the child's emotional state and adjust the game difficulty accordingly. The therapist would then primarily observe and monitor and would only intervene as needed in the otherwise automatic interaction.

**SO, IN THESE EXERCISES, IT WILL ALSO BE IMPORTANT FOR THE** ROBOT **TO RECOGNIZE THE CHILD'S EMOTIONS. HOW EXACTLY WILL THIS EMOTION RECOGNITION WORK?**

**M.SA.** The robot will recognize emotions in two ways: our colleagues at *Fraunhofer IIS* are working on visual emotion recognition via video data, and we are working with acoustic emotion recognition via audio data. The valence, that is, how negative or positive an emotion is, can be better detected via video, while the strength of an emotion is better detected via audio. For these reasons, we combine both.

**HOW WELL DOES THAT WORK WHEN DEALING WITH AUTISTIC CHILDREN?**

**M.SA.** So far, we have only fed our models with data from 147 participants, including 100 neurotypically developed children. However, we plan to improve our model with data from autistic children in the future. Pilot studies that Simone Kirst and the team at *HU-Berlin* conducted already clarified that autistic children do not like to speak as much or as often as neurotypically developed children. In such a case, we infer less from audio data and the video data becomes particularly important.

**S.K.** So far, the algorithms are still not adapted to autistic children. Therefore, it is currently very important that the therapist decides and judges the emotional expression of the child, and not the system.

**THE THERAPIST STILL PLAYS AN IMPORTANT ROLE, SO WHY IS A** ROBOT **NECESSARY TO SUPPORT THERAPY SESSIONS?**

**S.K.** A major advantage over a human therapist is the unlimited and step-by-step presentation of learning content presented in the most consistent form and structure. For autistic children this is of great value, as they need routine and consistency. In contrast to a robot, a human being cannot always present exercises in exactly the same way. The robot can also express and present learning content and language in a more simplified way. For children with autism who have partial language or cognitive impairments, it is much easier to follow the same sentences in simple language. Finally, there is the factor of motivation. Many autistic individuals have a special interest in technology and are fascinated by robots and as a result, they also have increased interest in learning content. We once had an autistic boy who rarely approached other children, but immediately talked to the robot after he came into the testing room.

**M.ST.** For autistic children, it is also extremely difficult to comply with social norms, such as maintaining eye contact or using polite phrases. When interacting with a robot, they are free from these stressors and aren't forced to fulfill social norms because the robot does not judge them. For example, if they don't understand their counterpart directly, or even slip in tone, nothing awful happens as the robot is psychologically invulnerable. This means that the children can experiment with the robot, or ask a question for the thirty-third time, without annoying it. Thus, the robot is less intimidating and anxiety-inducing than a human partner and creates a stress-free and relaxed learning atmosphere.

**IN COMPARISON, THEN, WHAT ARE THE THERAPIST'S STRENGTHS?**

**S.K.** The therapist basically serves as the opposite. The robot is ideal because it shows everything in a routine manner. However, the world is not always routine, which is why we need humans. Autistic children cannot automatically transfer specifically learned content to other settings or people by themselves. So, the therapist adds the human and the complexity factor to the child-robot interaction. Particularly, a reflection on an emotional situation is still easier with a human than with a robot, which only reacts to voice input and responds accordingly by means of a fixed program. However, the robot can provide the hook, and if the two are intelligently combined, both can provide their "superpowers" at the appropriate moment and complement each other ideally.

THERAPIST

ROUTINE

MARTINA SIMON, SIMONE KIRST, MARTIN STREHLER, JULIAN SESSNER, MILENKO SAPONJA

**SO FAR, WE HAVE MAINLY TALKED ABOUT THE STRENGTHS OF THE ROBOT AND THE THERAPIST. ARE THERE ALSO RISKS OF USING A ROBOT IN AUTISM THERAPY?**

M.SI.   We have worked out the risks together with therapists, parents of autistic children and adults on the autism spectrum during our workshops on ethical, social and legal aspects. On the one hand, there are data protection concerns. On the other hand, there are concerns that the robot could replace the therapist, that the child could develop an emotional bond with the robot, or that the child might learn robotic behaviors. So, there is a certain amount of skepticism, on which the acceptance of the robot in autism therapy hinges. Whether the robot will finally be accepted by stakeholders, that is, therapists and parents of autistic children as well as adults on the autism spectrum, is still under investigation.

**HOW DO YOU DEAL WITH THESE CONCERNS OF THERAPISTS AND PARENTS?**

S.K.   We've tried to *frame* the robot clearly as a tutor and not as a friend, so that the child will most likely not build an emotional bond with the robot that is too deep. We also plan to integrate an introductory story in which the robot verbalizes that it was programmed by humans and built in a factory.

**THERE ARE MANY DIFFERENT KINDS OF ROBOTS. WHY ARE YOU WORKING SPECIFICALLY WITH PEPPER?**

J.S.   A significant advantage was that *Pepper* is already a finished product. Thanks to the software supplied by the manufacturer, rudimentary functions of the robot are already available, so we were able to implement initial behavior and carry out initial studies relatively quickly. By using additional external software solutions, however, we are also quite flexible with regard to other robots.

M.ST.   Actually, we considered 7-8 robots. *Pepper*, however, was the most suitable for our case. On the one hand, it is about the same size as the children, so they communicate almost at eye level. On the other hand, it has no facial expresions itself, but has the tablet on which we can show real people's faces.

S.K.   Due to its size and humanoid appearance, it probably appears more convincing and serious in its tutor role than the small *Nao* or non-humanoid robots. By using its tablet, it can present and explain learning content like a teacher would on their

blackboard. Besides, *Pepper* has a very differentiated gesture and body language, which can be used well for relaxation exercises for example. Children and parents surveyed in our initial usability and acceptance studies rated *Pepper* in our modelled tutor role as intelligent, sympathetic and empathetic—characteristics that fit a good tutor.

**SO, WHAT EXACTLY IS THE ADVANTAGE OF THE ROBOT OVER A TABLET APPLICATION?**

S.K.   We have just planned a study comparing tablet applications and the robot to find out exactly that. I think the physical exercises, like the relaxation exercise, can be done better with the robot because of its embodiment. Other learning content, like emotion recognition, could probably work just as well with the tablet. For further work, we will have to carefully consider the purpose for which the physical part of the robot is needed: Where exactly is the added value compared to the tablet?

NEITHER HUMAN NOR COMPUTER

MARTINA SIMON, SIMONE KIRST, MARTIN STREHLER, JULIAN SESSNER, MILENKO SAPONJA

Inspired by the interview, the illustrator Johanna Benz visually commented on chances, risks and scenarios of robotics from her own perspective. © Johanna Benz

# Dr. Lenneke Kuijer

is Assistant Professor in the *Future Everyday Group* in the *Department of Industrial Design* at the *TU Eindhoven (NL)*. Her research combines sociology and design theory to better understand the role of interactive technologies and their designers in the dynamics of everyday life, particularly around energy demand in the home.

# Counting Characters and Spaces—On Robot Disabilities, Robot Care, and Technological Dependencies

Lenneke Kuijer

I want to begin this commentary by congratulating the editors on their critical, innovative collection. It contains thought-provoking examples and outcomes, many of which are directly applicable for designers. *[ehm, 318 characters and spaces]*

From the perspective of my research into the secondary effects of technology design on everyday life, a couple of critical reflections and questions arise. This begins with a definition. What is a social robot? In my view, a robot is a device that semi-autonomously executes tasks. This is possible due to relatively new technological capabilities, such as (battery) power, sensors, actuators, processors, and connectivity. In this definition, a smartphone is not a robot because it doesn't have actuators, but a washing machine is. Social, to me, implies robots that participate in everyday life rather than operate somewhere out of sight. *[around 950 characters and spaces, about 5,000 left to the limit given by the editors]*

## ROBOT **DISABILITIES**

The editors' position that social robots shouldn't be designed to imitate humans is something I fully agree with; robots have their own unique strengths and weaknesses. However, when reading the contributions to this book, what struck me was that there is ample attention to the unique *strengths* of robots, referred to as "superpowers" (→ p. 44), but much less to their *weaknesses*. Robot "disabilities," such as responding to unexpected situations or behaving appropriately in socially complex ones, are mentioned, but not reflected on in depth. These disabilities imply that in complex social contexts such as homes, supermarkets and hospitals, robots are bound to act inappropriately in some cases, simply because not all situations they end up in can be predicted. People engaging with them in these moments have little choice but to accommodate this behavior. *[this counting of characters and spaces is too tedious]*

Moreover, this responsibility to deal with robot disabilities might not be carried equally. Service robots take up space and will not be cheap. It can therefore be expected that their services are more accessible for people with larger homes and higher incomes. Neighborhood shops are part of the glue of a community. If certain people visit shops less, this could further limit the bubble they live in already, while the people who do visit and work in local shops have to deal with robots—which they do not own and did not design—that will need assistance when something is out of stock, a lane is blocked, or shelves are knocked down.

### ROBOT CARE

This issue sits within broader issues of robot care. Robots, more than artifacts with less complicated behavior, need humans to function properly. Think for example of the robot vacuum cleaner, which needs humans to tidy the room, to empty its bin and keep it clean. Service robots need care, or their *lives* will be short. Who is going to care for the robots? I believe the Techno-Mimesis approach (→ p. 140) is suitable for identifying some aspects of these robot needs (→ p. 78), but it could be developed further to identify longer-term needs related to maintenance.

### TECHNOLOGICAL DEPENDENCIES

But the bigger issue in my view lies in what robots *can* do. Yes, robots have superpowers; they are capable of things that humans cannot do, or do not do as well, such as endless patience, or sucking dirt from a floor. Once these unique capabilities find their place in everyday life, they become difficult to remove. Think, for example, of getting rid of washing machines—uniquely capable of spinning a drum at 1400 rounds per minute, and never complaining about doing the laundry at night.

A unique quality of humans is that they are *multi-purpose*. One human can clean, shop, and care, and on top of this gossip, walk, smile, and do much more. Robots tend to be more single-purpose. Delegating different kinds of human work to robots therefore requires a growing army of devices: dishwashers, robot vacuum cleaners, washing machines, thermostats. This has clear benefits, but robots require energy and materials to produce and transport, electricity to function and communicate, and space to *live* and operate. When designing new functionalities and roles for robots, I believe we should ask what kind of dependencies this technology is creating, and at what cost—if we get used to robots caring for our elderly and doing our shopping, is there a way back? *[around 1000 characters and spaces left …]*

To meet the terms of the *Paris Agreement*, we need strongly reduced ecological footprints. While this will partly be achieved through new technologies, the *European Environment Agency* reports that, over recent decades, "half the efficiency gains achieved through technological innovation in the household sector were offset by the increasing number of electrical appliances and by larger homes" (European Environment Agency, 2019). Who counts the number of artificial *characters* and the spaces they require in everyday life?

I believe designers can play a role in setting and practicing limits here. In my teaching, I use an image of a *Netflix*-watching robot to trigger reflections on which conveniences are really needed. Moreover, sensible design decisions can make huge differences: why have a fully fledged robot shop in a supermarket when shopping online is a much less resource-intensive way to restock? And, I have to admit, there can be positive side effects from delegating tasks to robots that can be exploited. For example, robots are less prone than humans to impulse buying, and better at counting.

*P.S. I couldn't have written this piece without the capability of my computer to count characters and spaces. However, if my computer had not had this function, then the editors would probably have indicated the maximum length of this piece in another—more "humane"—metric. Artificial capabilities are intertwined with everyday life, but some we can do without. Whether we should, or whether we want to—in this case—is another question.*

European Environment Agency (2019). Progress on energy efficiency in Europe. Retrieved May 4, 2021, https://www.eea.europa.eu/data-and-maps/indicators/progress-on-energy-efficiency-in-europe-3/assessment.

# Designing Robots with Personality

Lara Christoforakos
Sarah Diefenbach
Daniel Ullrich

Let's imagine a time when cleaning robots have become very popular, and every household has its own. Whether this cleaning robot is humorous, sociable, and sometimes makes cocky jokes, or is rather depressed and sarcastic and avoids actively seeking interaction with humans, has different implications for users. It would certainly affect users' mood and behavior in different ways, just like the different personalities of our human interaction partners affect us, each in a different but characteristic way. To foster responsible robot design, such effects on users need to be considered as part of the overall intended impression a robot should leave. This goal has been developed in many of the robotics projects with which we have co-operated during the last two years.

## WHY SHOULD A ROBOT HAVE A PERSONALITY?

First of all, the question arises as to why a robot should have a particular type of personality, and what benefits this could provide. In current human-robot interaction (HRI) research, the definition and operationalization of personality in robots is the subject of much research and ongoing debate. On the one hand, many researchers argue that designing a robot with a personality, and thus mimicking qualities known from humans or animals, might not always be ideal. For example, Laschke et al. (2020) suggest that mimicking humans in robot design could reinforce inappropriate gender stereotypes (Brahnam and De Angeli, 2012), or affect children's behavior in as-yet unknown ways (Sciuto et al., 2018). Moreover, in the private home context companion technologies are typically involved in intimate situations, including interactions with household members. With regards to data protection and the desire for privacy, some users might prefer a technology with less social cues (see, e.g., Ha et al., 2020), i.e., one that does not have a personality and does not resemble any human counterpart.

LARA CHRISTOFORAKOS, SARAH DIEFENBACH, DANIEL ULLRICH

On the other hand, robots are increasingly being applied to address users' social needs. For example, within the domain of mental health or elderly care, social robots are often implemented to support users' wellbeing by enhancing social interaction. In these cases, it may be a reasonable or even necessary goal to design robots that mimic human qualities, e.g., with certain personality traits. This could come with the advantage of reinforcing intuitive interactions known from human-human interactions (see, e.g., Laschke et al., 2020). Moreover, for innovative visions of robots as roommates, it can foster user acceptance to design robots that behave in a more unpredictable and independent manner rather than submissive robots that depend solely on their users' commands (c.f., Auger, 2014). As a result, robots are designed with certain personality traits.

# WHAT TYPE OF PERSONALITY SHOULD A ROBOT HAVE, AND HOW CAN THIS BE IMPLEMENTED?

Having decided to equip a robot with a certain personality, the next question concerns what type of personality a robot should have, and how this can be expressed through design. Specifically, practitioners have to reflect on how the robot will interact with users. Should it be supportive and understanding, or demanding and rather strict? Often, practitioners also express the goal of developing an adaptive personality suitable for various users and situations. This goal was part of several of the robotics projects presented and discussed in this book. This in turn brings up the question of how the robot's personality should be designed depending on the user and specific interaction situation.

To date, universal approaches for implementing a desired robot personality do not exist. In general, previous research on social robot personality highlights challenges in the definition of robot personality rather than obvious solutions. For example, attempts to systematically conceptualize robot personality based on approaches from →personality psychology reveal many inconsistencies and barriers involved in transferring models of interpersonal interaction to the domain of robots (for a comprehensive overview, see Diefenbach et al., forthcoming). Moreover, after discussions and consideration of various design solutions within many of the projects we cooperated with, focusing on the overall human-robot relationship and the respective role of robots within specific contextualized interactions appears more practicable than focusing solely on the design and implementation of a robot's personality in isolation. In parallel to most of the abovementioned projects, social robots are often applied in the

**Personality Psychology** is a central field in psychology. It involves research on the psychological concept of personality and how this varies among individuals (see, e.g., Costa & McCrae, 2011).

context of healthcare or private homes. In different situations, such as interactions with various users like patients and nursing staff, different robot behaviors or styles of communication were shown to be beneficial (c.f., Niess and Diefenbach, 2016). For example, while rehabilitation patients might benefit from a motivational robotic counterpart which is highly present and proposes many activities, nursing staff might need a more neutral counterpart or supportive assistant. Therefore, an isolated consideration of robot personality appears to be less expedient, as the robot typically acts in interaction with its user. Consequently, it seems more appropriate to focus comprehensively on the human-robot relationship and the specific roles of robots and humans in interactions.

Examples of different roles of robots and humans in interaction can be found in preliminary frameworks in the HRI literature. In their review of HRI frameworks, for example, Onnasch and Roesler (2020) have developed a new HRI taxonomy that considers the human, robot, interaction, and HRI context. Within this framework, the authors also specify human roles that the robot can adapt in specific HRIs. These roles involve the supervisor, operator, collaborator, cooperator, and bystander.

Furthermore, specifically considering →companion technologies as a form of technological counterpart in business contexts, Niess et al. (2018) have identified two basic types of companions: active and passive. When reflecting on their personal experience of companion technologies, study participants typically characterized an active companion as innovative, dominant, proactive, and independent; some also reported a feeling of being under surveillance and limited in their autonomy. A passive companion was described as caring, empathetic, cautious, subdominant, and only acting on direct request (Niess et al., 2018). Furthermore, beyond the active-passive distinction, the authors also found a variety of possible roles and character traits associated with the image of a companion. According to their results, a (digital) companion can take the role of a friend, advisor, teacher, or coach, while each of these roles was connected to different expectations, as well as requirements for interaction qualities towards the product. The authors give detailed examples for patterns of interaction qualities related to different companion roles resulting from the workshops conducted (c.f., Niess et al., 2018). For example, a participant seeking motivation to train in order to look good on a beach vacation deliberately decided in favor of a tough communication style of the digital companion in the used fitness app to achieve this goal as fast as possible.

Moreover, considering embedded technologies in the smart home context, Diefenbach et al. (2020) propose possible roles for what they call a "room intelligence" each displaying certain characteristics (→Fig. 1). The authors propose an overarching interaction

**Companion Technologies** can be described as interactive technological artifacts that evoke empathy (Niess & Woźniak, 2020).

concept for a whole environment, which conveys the mental model of a central, omnipresent, and embodied room intelligence. Although such technologies are described as having "personalities" (c.f., Diefenbach, 2020), they are designed to adopt specific roles in interactions with users or their environment rather than to express an isolated, consistent personality based on dimensions from personality psychology.

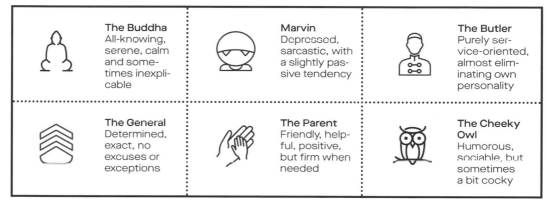

**Fig.1** Possible personalities for a "room intelligence" (Diefenbach et al., 2020).

In general, it appears to be common and practicable in human-computer interaction (HCI) and HRI research to consider specific roles for certain interactions between humans and technology. The frameworks and specific roles mentioned above can serve as a basis for the design of robots as social counterparts, as well as for choices regarding their roles within a specific HRI. In line with this, some of the projects presented in this book describe successful examples of robot design achieved through focusing on a specific role in HRI, e.g., a friend or coach, depending on the specific interactions required within a given household.

## HOW CAN A ROBOT PERSONALITY BE EVALUATED?

Finally, it is essential to evaluate how users perceive the robot. While the designer may have intended to apply a very supportive and service-oriented robot personality, users might perceive the personality to be, for example, ignorant and distant. This, in turn, could affect the user experience in an unintended way. Therefore, potential prototypes should be evaluated in terms of perceived robot personality, among other considerations.

In current research, assessment methods for personality perception are diverse, but are not necessarily appropriate for certain contexts or robots. Additionally, the selection of evaluated user perceptions does not follow a particular theoretical approach due to a lack of comparability (for a comprehensive overview, see Diefenbach et al., forthcoming). In this regard, in our latest research, we have developed a tool, the so-called Robot Impression Inventory (RII). The inventory consists of a set of dimensions (e.g., appearance, movement, personality) of users' impressions of robots, each consisting of various facets (e.g., emotional stability, emotional vulnerability, openness of the personality dimension) (Ullrich et al., 2020).

The RII can be used as a source of design inspiration, for example, by exploring different facets of a dimension that could be addressed through design. The inventory could also be applied as a questionnaire and evaluation tool, for example, to test whether an intended robot personality is perceived accordingly by potential users. Furthermore, as the inventory involves different dimensions, empirical interrelations between dimensions could be explored. Results could, for example, indicate how visual design cues, primarily affecting the perceived appearance on the robot, could be interrelated with the perception of various personality facets. The RII was validated within a set of studies and applied in preliminary evaluation studies of the eight robotic projects discussed in this book.

## WHAT REMAINS TO BE EXPLORED IN ROBOT PERSONALITY DESIGN?

Future research could aim at a more systematic categorization and definition of roles for robotic counterparts regarding various forms of human-robot relationships, e.g., companion robots and human-robot teaming, as well as automation. In addition, future studies could explore how the defined roles affect various user experience variables of interest, such as acceptance, trust, and overall evaluation of the robots, as initial studies have done for single role comparisons (see, e.g., De Graaf and Alouch, 2015; Griffiths et al., 2021; Groom and Filipe, 2008). The robot's specific operational context should be considered, as this could have an impact on such evaluations, for example, users might have different expectations of a cleaning robot versus a robot coworker in the office. Based on such results, suggestions for appropriate robot roles within specific HRIs could be compiled, and design ideas to facilitate the incorporation of appropriate roles could be developed and evaluated.

More research is also needed on the interrelation between robot roles and user personalities, as well as how robots can adapt to users depending on their individual preferences and personalities—the idea of adaptable robot roles implies that robots will act differently depending on context and their interaction counterpart. In many contexts where social robots are used, such as care homes or private households, this could imply that one user might observe the robot behaving in a completely different way while interacting with another user. This raises the question of whether and how this could affect robot authenticity and thus the overall user experience.

In summary, especially within the field of social robots, the idea of designing robots with personalities has gained increasing interest, and this can be beneficial when a robot's main purpose is social interaction with its users. The systematic conceptualization and implementation of a coherent robot personality relying on theoretical approaches from personality psychology appears rather challenging. Instead, our cooperation with robot designers in different contexts suggest focusing on robot roles in terms of specific interactions between humans and robots. The abovementioned frameworks offer preliminary ideas for orientation in this regard. Thus, tools such as the RII can support a systematic process of robot personality design through an assessment of users' impressions. Future work should address an overarching classification of robot roles, further considering specific tasks and interaction contexts. Such studies could then serve as a basis to advance research and development of appropriate robot roles for specific HRIs, and to develop concrete design solutions.

Auger, J. (2014). Living with robots: A speculative design approach. *Journal of Human-Robot Interaction, 3*(1), 20–42. https://doi.org/10.5898/JHRI.3.1.Auger

Brahnam, S., & De Angeli, A. (2012). Gender affordances of conversational agents. *Interacting with Computers, 24(3)*, 139–153. https://doi.org/10.1016/j.intcom.2012.05.001

Costa, P. T., & McCrae, R. R. (2011). The five-factor model, five-factor theory, and interpersonal psychology. In L. M. Horowitz & S. Strack (Eds.), *Handbook of interpersonal psychology: Theory, research, assessment, and therapeutic interventions* (pp. 91–104). John Wiley & Sons, Inc.

De Graaf, M. M., & Allouch, S. B. (2015). The evaluation of different roles for domestic social robots. In Z. Wang & K. Wada (Eds.), *2015 24th IEEE International Symposium on Robot and Human Interactive Communication* (pp. 676–681). IEEE Press. https://doi.org/10.1109/RO-MAN35437.2015

Diefenbach, S., Butz, A., & Ullrich, D. (2020). Intelligence comes from within—personality as a UI paradigm for smart spaces. *Designs, 4*(3). https://doi.org/10.3390/designs4030018

Diefenbach, S., Herzog, M., Ullrich, D., & Christoforakos, L. (forthcoming). Social robot personality: A review and research agenda. In C. Misselhorn, P. Poljanšek & T. Störzinger (Eds.): *Emotional machines: Perspectives, affective computing and emotional human-machine interaction.* Springer.

Griffiths, S., Alpay, T., Sutherland, A., Kerzel, M., Eppe, M., Strahl, E., & Wermter, S. (2021). Exercise with social robots: Companion or coach? In S. Schneider, S. S. Griffiths, C. A. Cifuentes G., S. Wermter & B. Wrede (Eds.), *Proceedings of Workshop on Personal Robots for Exercising and Coaching at the HRI 2018.* Association for Computing Machinery.

Groom, V., & Filipe, J. (2008). What's the best role for a robot. In J. Andrade Cetto, J.-L. Ferrier & J. Filipe (Eds.), *Proceedings of the International Conference on Informatics in Control, Automation and Robotics* (pp. 323–328). Springer.

Ha, Q. A., Chen, J. V., Uy, H. U., & Capistrano, E. P. (2020). Exploring the privacy concerns in using intelligent virtual assistants under perspectives of information sensitivity and anthropomorphism. *International Journal of Human–Computer Interaction, 37*(6), 512–527. https://doi.org/10.1080/10447318.2020.1834728

Laschke, M., Neuhaus, R., Dörrenbächer, J., Hassenzahl, M., Wulf, V., Rosenthal-von der Pütten, A., Borchers, J., & Boll, S. (2020). Otherware needs otherness: Understanding and designing artificial counterparts. In D. Lamas & H. Sarapuu (Eds.), *Proceedings of the 11th Nordic Conference on Human-Computer Interaction: Shaping Experiences, Shaping Society* (pp. 1–4). Association for Computing Machinery. https://doi.org/10.1145/3419249.3420079

Niess, J., & Diefenbach, S. (2016). Communication styles of interactive tools for self-improvement. *Psychology of Well-Being, 6*(3), 1–15. https://doi.org/10.1186/s13612-016-0040-8

Niess, J., Diefenbach, S., & Platz, A. (2018). Moving beyond assistance: Psychological qualities of digital companions. In T. Bratteteig & F. R. Sandnes (Eds.), *Proceedings of the 10th Nordic Conference on Human-Computer Interaction* (pp. 916–921). Association for Computing Machinery. https://doi.org/10.1145/3240167.3240240

Niess, J., & Woźniak, P. W. (2020). Embracing companion technologies. In D. Lamas & H. Sarapuu (Eds.), *Proceedings of the 11th Nordic Conference on Human-Computer Interaction: Shaping Experiences, Shaping Society* (pp. 1-11). Association for Computing Machinery. https://doi.org/10.1145/3419249.3420134

Onnasch, L., & Roesler, E. (2020). A taxonomy to structure and analyze human-robot interaction. *International Journal of Social Robotics, 13*(4), 833–849. https://doi.org/10.1007/s12369-020-00666-5

Sciuto, A., Saini, A., Forlizzi, J., & Hong, J. I. (2018). "Hey Alexa, what's up?" A mixed-methods studies of in-home conversational agent usage. In I. Koskinen & Y.-K. Lim (Eds.), *Proceedings of the 2018 Designing Interactive Systems Conference* (pp. 857–868). Association for Computing Machinery. https://doi.org/10.1145/3196709.3196772

Ullrich, D., Diefenbach, S., & Christoforakos, L. (2020). Das Roboł Impression Inventory—Ein modulares Instrument zur Erfassung des subjektiven Eindrucks von Robotern. In T. Köhler, E. Schoop & N. Kahnwald (Eds.), *Gemeinschaften in Neuen Medien. Von hybriden Realitäten zu hybriden Gemeinschaften: 23. Workshop GeNeMe'20* (pp. 244–249). TUDpress.

IMPULSES AND TOOLS

DESIGNING ROBOTS WITH PERSONALITY

LARA CHRISTOFORAKOS, SARAH DIEFENBACH, DANIEL ULLRICH

# VIVA

GUESS!

## AIM OF OUR RESEARCH PROJECT

The aim of the project is the development and design of a social robot. The robot is meant to improve the emotional state of the user and earn their trust by means of a vivid presence and sympathetic interaction. *VIVA*, the robot, thus becomes an attentive and empathetic flatmate.

## CONTEXT, ROLE AND TASK OF OUR ROBOT

The robot *VIVA* is a flatmate that provides companionship and improves the wellbeing of the user. It can be used in private households as well as in care homes for elderly people.

## MOST ASTONISHING FINDING LEARNED ABOUT ROBOTS DURING OUR WORK

The reaction to *VIVA* is mostly emotional. As a matter of principle, some people reject such social robots while others are totally enthusiastic. At the same time, however, it has been shown that most people have ambivalent feelings about social robots.

## WHO WE ARE

University of Bielefeld, Social Cognitive Systems and Social Interaction; University of Bielefeld, Applied Social Psychology and Gender Research; University of Augsburg, Human-Centered Artificial Intelligence; University of Applied Sciences of Bielefeld, Commercial Law; Neuland Software GmbH; Visions4IT GmbH; Navel Robotics GmbH.

# Designing Robots as Social Counterparts—A Discussion about a Technology Claiming its Own Needs

*Interview with*

Claude Toussaint (C.T.)
Sonja Stange (S.S.)
Julia Stapels (J.S.)

*By*

Lara Christoforakos
Tobias Störzinger

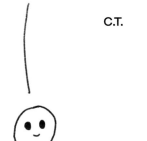

**HOW WOULD YOU DESCRIBE VIVA AS A CHARACTER AND HOW DID YOU DEFINE AND DEVELOP SUCH A CONCEPT?**

C.T.  We have always had the idea of an animated 3D comic figure, similar to those in *Disney* or *Pixar* movies, combined with the image of a pet in mind. The early communication of this image within our team was very important. We discussed different scenarios and tried to specify *VIVA*'s reactions in order to establish a consistent image of the robot. From there, we derived specific needs, emotions and resulting general behaviors. It has proven important for the perception of character that the robot is perceived to be autonomous. It should not only react to the user, but rather behave and interact proactively. Generally, to design a coherent character, we included subjective experiences as well as designers' and developers' expertise, but also followed certain systematic steps, for example, theoretical frameworks in psychology.

**WHAT KIND OF ROLE WOULD YOU ASSIGN TO VIVA DURING THE INTERACTION WITH ITS USER?**

C.T.    *VIVA* represents something between a coach and a friend. Considering the personality types proposed by Christoforakos and colleagues (→ p. 70), a combination of the "Buddha" as an "all-knowing, serene, calm and sometimes inexplicable" personality and the "cheeky owl" as a "humorous, sociable but sometimes a bit cocky" personality could be appropriate. Still, the theme of humor and fun should be added. These two types appear lively and cheeky, which could be a general goal in the robot's character design.

**FOR THE DESIGN OF ITS CHARACTER, YOU MENTIONED SPECIFYING NEEDS FOR VIVA. WHY SHOULD A ROBOT GIVE THE IMPRESSION OF HAVING NEEDS?**

S.S.    We decided early on that *VIVA* should have its own needs, as this fosters the perception of it having a personality. These needs form the basis of *VIVA*'s actions, but still *VIVA* also considers the user's needs. *VIVA* has mechanical needs, such as needing to be charged or wound down when overheated, as well as social needs, such as social contact, entertainment, and certainty, that is, the need to explore unexpected stimuli. On the other hand, the robot should be able to recognize the user's needs for social contact and relaxation. To satisfy both the user's and its own needs, *VIVA* selects appropriate behavioral strategies. The weighing of its own needs, as well as the consideration of the user's needs, leads to the appearance of *VIVA*'s personality.

**HOW DOES THE WEIGHING OF NEEDS HAPPEN? SHOULD VIVA ALWAYS SATISFY THE USER'S NEEDS?**

S.S.    The user's needs regulate *VIVA*'s behavior. Let's say *VIVA* has a high need for social contact but the user doesn't. *VIVA* would realize this and choose a happy medium to satisfy its own needs, such as carefully testing whether the user wants company instead of simply heading towards them. *VIVA* thus tries to find parallel solutions to target both its own and the user's needs. In the end, our goal is that *VIVA* adjusts its behavior and resulting character to the user's preferences. *VIVA* may be socially intelligent, but the weighting of different needs, and therefore how empathetic, benevolent or self-centered *VIVA* is, should be customizable by the user.

**WHAT HAPPENS WHEN VIVA'S NEEDS REMAIN UNSATISFIED?**

**S.S.**  *VIVA*'s emotions depend on its needs, so when its needs cannot be satisfied, *VIVA*'s mood and emotions are affected.

**WHICH EMOTIONS CAN VIVA CONVEY?**

**C.T.**  We work with the continuous model of Valence, Arousal and Certainty.

**VIVA'S EMOTIONS COULD AFFECT THE USER, COULDN'T THEY? WHAT CHALLENGES DO YOU EXPECT TO COME WITH THAT?**

**J.S.**  They certainly can. We decided to permit *VIVA* to act unexpectedly, in order to illicit unanticipated reactions aimed at increasing long-term acceptance. Research shows that attitudes towards social robots are inherently ambivalent. There are even negative emotions and a reluctance to have a robot in the home.

**S.S.**  One possible reason for failed long-term interactions might be that robots are not perceived as *lively* enough but rather as submissive and lacking useful functionality. Our approach is to make the robot seem more *alive* through unexpected actions, like leaving the room when in need of relaxation. We hope that this way, the robot will remain an interesting interaction partner and foster the user's long-term acceptance. Of course, such actions might also evoke negative user emotions. We aim to address this challenge by means of transparency; the robot should be able to explain its actions, for example, "I needed to leave to recharge my battery." With this transparency of actions, we hope to strengthen the bond between robot and user. In the long run, the user should understand how the robot behaves and makes decisions. This should further enable the user to express purposive wishes regarding the robot's behavior, for example, "I don't want you to play music in the mornings." In line with these wishes, the robot's needs and resulting strategies could be weighed. It is our explicit goal to increase the user's well-being, on the one hand, by means of resonance, for example, mirroring of user's emotions through *VIVA*, and on the other hand, through specific interventions to improve personal wellbeing, such as interventions from positive psychology.

**SO, VIVA SHOWS A RANGE OF EMOTIONS THAT ARE DIRECTED AT A POSITIVE RELATIONSHIP. NOW, WHEN VIVA SHOWS NEGATIVE EMOTIONS—FOR EXAMPLE, ANGER OR SADNESS BECAUSE IT IS ALONE—USERS COULD FEEL THAT THEY OWE VIVA SOMETHING. DO WE OWE ROBOTS ANYTHING?**

C.T.     We have dealt with this in ethics workshops and—for the time being—we have not included negative emotions, such as anger, in *VIVA*. But of course, there are different opinions on such a complex topic, even within the team. Overall, we don't like the wording that *VIVA demands something*. *VIVA* should be able to be taken care of, because taking care of someone can enhance the user's wellbeing but caring for it shouldn't be demanding. If we make a comparison with a pet, we would say that with an animal, you have a moral obligation to be responsible for it, but with a robot we don't see it that way. With animals, one has to provide them with food and water for example. However, with *VIVA*, it would be a pity, and maybe also unwise, if one didn't allow *VIVA* to reach the charging station on its own, but it wouldn't be morally problematic. After all, *VIVA* has no consciousness, and even if people feel responsible for *VIVA*, that is like feeling guilty for not watering their plants. We therefore think *VIVA* should convey to users, for example, that it has no problem being alone for a long time. When users come home from work and then want to go out and meet friends, they shouldn't feel that they are neglecting *VIVA*. Instead, *VIVA* could happily say what it did while the user was away because *VIVA* is supposed to promote social contact, not prevent it by acting as a social competitor. In general, we aim to induce positive feelings through the robot and avoid inducing negative ones, such as guilt.

**VIVA SHOWS EMOTIONS AND VARIOUS NEEDS, SUCH AS SOCIAL CONTACT, BUT AS YOU SAY YOURSELF, IT HAS NO CONSCIOUSNESS AND DOES NOT FEEL EMOTIONS. DOESN'T VIVA DECEIVE USERS?**

C.T.     We aim to communicate only inner states that are incorporated in *VIVA*'s behavioral architecture and that actually influence its behavior, such as its robotic needs and emotions. Even though these may not be comparable to human needs or emotions, they are decisive for *VIVA*'s behavior. Some users might attribute phenomenal emotions to *VIVA*, but these users are deceiving themselves and are not actively deceived by *VIVA*. We see this as analogous to when people anthropomorphize their dog and believe, for example, its emotions are human-like. Anthropomorphism is an individual variable, thus different users might have different tendencies to attribute thoughts and feelings to the robot. Either way, non-verbal communication and the expression of emotions are—both with the dog and with *VIVA*—very helpful for interaction. Through emotions, the respective system can describe itself, and this is helpful for joint coordination. Emotions are also important for our goal of creating social resonance with *VIVA*; if my counterpart reflects my emotions, I feel seen.

**NOW, IF VIVA PROVIDES SOCIAL RESONANCE AND IS MEANT TO BE A COMPANION, CAN YOU HAVE ANY TYPE OF SOCIAL RELATIONSHIP WITH VIVA? OR ARE THERE SPECIFIC TYPES OF RELATIONSHIPS—A DEEP FRIENDSHIP, FOR EXAMPLE— THAT ARE NOT APPROPRIATE?**

C.T.  First, we find it problematic to use terms that deal with interpersonal relationships. *VIVA* is not a human being and is not intended to replace one. Therefore, terms like *friendship* are very misleading here. This begs the fundamental question about how *appropriate* human-robot-relationships can be developed and what *appropriate* even means in this context. If the relationship helps people and increases their wellbeing, why should there be anything negative about that? Of course, the relationship with *VIVA* should not be exploited and misused. But this, of course, depends a lot on how people will deal with this technique and how they will use it.

THIS BRINGS US BACK TO THE BEGINNING: MANY SOCIAL RELATIONSHIPS ARE CHARACTERIZED BY HAVING TO NEGOTIATE HOW TO RELATE TO EACH OTHER. WOULD YOU SAY THAT YOU CAN ALSO HAVE THIS FORM OF COEXISTENCE WITH VIVA? SHOULD WE LET VIVA HAVE A SAY IN THIS PROCESS?

C.T.   Of course, negotiation with robots could be problematic if they are too dominant or even manipulate users. But in the case of *VIVA*, again it is comparable to the interaction with pets: there are people who live their lives according to their dogs, which can be problematic. To prevent manipulation, there are certain practices we do not want for *VIVA*. For example, when people want advice from *VIVA*, we believe that *VIVA* can provide only a certain amount of coaching, like with generic follow-up questions that encourage reflection. However, *VIVA* should not give an opinion on concrete proposals for action, as a human might be able to; *VIVA* may be able to help you find solutions yourself, but we wouldn't want a robot to give concrete advice.

Inspired by the interview, the illustrator Johanna Benz visually commented on chances, risks and scenarios of robotics from her own perspective. © Johanna Benz

## Brigitta Haberland

is a Pastoral Counselor at the *Protestant University in Rhineland-Westphalia-Lippe*, and a Psychosocial Counselor at the *University of Health in Bochum*. After studying Protestant theology and social pedagogy, she worked as a Project Manager in the field of school and social inclusion of people with disabilities at a church agency.

## Dr. Karsten Wendland

is Professor of Media Informatics at *Aalen University* and Senior Researcher at the *KIT*. He studied computer science at *Darmstadt University of Technology*, was a research associate at the *Center for Interdisciplinary Technology Research (ZIT)* and earned his doctorate in technology design.

## Dr. Janina Loh

is an Ethicist (Stabsstelle Ethik) at *Stiftung Liebenau in Mecken-beuren* on Lake Constance. They received their doctorate at the *Humboldt University in Berlin.* Their narrower research interests include responsibility, trans- and posthumanism, and robot ethics, as well as Hannah Arendt, feminist philosophy of technology, theories of judgment, and ethics in the sciences.

# Falling in Love with a Machine—What Happens if the Only Affection a Person Gets is from Machines?

*Moderated by*

Brigitta Haberland (B.H.)
Karsten Wendland (K.W.)
Janina Loh (J.L.)

Felix Carros
Anne Wierling
Adrian Preussner

**IF** ROBOTS **INCREASINGLY EXIST IN OUR EVERYDAY LIVES AND EXHIBIT SOCIAL BEHAVIOR, COULD** HUMANS **DEVELOP FEELINGS FOR** ROBOTS**?**

J.L.     There's no question that they can, because we're already doing that today. There are people who officially say that they are in love relationships with objects. There is also an official name for this: "objectophilia." Originally, this term actually had a positive connotation, as it has been used by people as a self-attribution. In one case, there was a person that had relationship with a model airplane. In addition, there have been various other cases in the USA where people have married their smartphones, or of people in Asian metropolitan areas who have had relationships with avatars. What is ethically interesting and questionable here is whether a company deliberately sells such avatars so that people can have a relationship with them. In any case, many examples exist already where people have fallen in love with diverse objects. Some of these objects are box-shaped, such as smartphones. But others have no humanoid-like shape as well, such as model airplanes or the Berlin Wall. Contrast that with humanoid figures like avatars, which are modeled after fictional singers, or humans. That's why I believe that we as a human species can theoretically fall in love with all kinds of forms, no matter what shape they have.

> **BUT IT MAKES A DIFFERENCE WHETHER A HUMANOID ROBOT DESIGN OR A ROBOT-BOX DESIGN IS USED. THAT ALSO ENTAILS A CERTAIN KIND OF MANIPULATION. IS THAT SOMETHING I SHOULD DO AS A DEVELOPER OF SUCH ROBOTS?**

J.L.  The accusation that this is a form of manipulation has been discussed for more than ten years, especially in the case of care robots, such as *Paro or Pepper* (→ Figs. 3 and 4, p. 33 and 34). The question often arises whether it violates the dignity of the person when a robot pretends to have feelings. Robots fake feelings just like human actors and other artificial figures. Perhaps one can accuse the manufacturer of such robots of deception. But that's different from saying that a robot such as *Pepper or Paro* is trying to deceive us.

B.H.  I agree, the robot itself cannot manipulate me. I see the danger of manipulation more on the part of those who make the device. Through these devices there are other possibilities of manipulation. Unlike with humans—who, of course, also manipulate when communicating—with robots, the triggers for manipulation are mathematically predictable.

K.W.  Intuitively, one would first say that this is totally naïve, to replace a human with a robot. If a care home resident develops an affection for a device, this device does not develop the sympathies back, regardless of whether it is a black box or a pretty humanoid. The *objectophilia* is directed in one direction; the Berlin Wall will not love a person back. So, beware of self-absorption and fantasy worlds! Another topic is incapacitation, where a resident's situation is exploited to keep them quiet. Who would do that? I could imagine that some overworked care workers would accept it because they can organize their work better if the residents are kept busy. One level higher, management also thinks about costs. Does the care robot pay off? It has no entitlement to holidays or salary increases, so such facilities could possibly work much more economically. In this respect, I observe that many robotics developers are extremely clueless, almost naïve, and often only consider the immediate aspects, completely free of context.

**J.L.**    I would agree with the vast majority of the points you made. In principle, I would say that there are two sets of ethical questions here. One is the question of what ethical challenges we face from the object itself. How can someone be manipulated, instrumentalized or misappropriated? After all, robots are being developed for assistance, not to replace caregiving activities. The other is about social interaction. The interactions are not real because they are not symmetrical and equal. A social interaction has a value in itself due to the counterpart's partial unavailability and uncontrollability. Unlike a robot, we can't put our human counterpart aside. As long as it is ensured that interaction with a robot is not intended to replace other (and *real*) social interaction, there is no reason why interaction with a robot should be problematic. Especially since in the care sector interaction with social robots, like *Paro*, can also have positive effects on human-human interaction between the resident and caregiver. Here I see no arguments why the interaction with a robot should be excluded in advance, but I find your (Karsten Wendland) objections understandable. I think these are just forms of misuse and abuse of such technology.

**K.W.**    What the robot should look like—on this question I would like to reinforce the aspect of inaccessibility mentioned by Janina Loh—plays an essential role. For all the humanity a robot can exude, its product design should have a clearly recognizable on/off button. Manufacturers are aware of the possible seductive qualities of products, so that smartphones, for example, are deliberately designed to be elegant so that they are caressed. Marc Hassenzahl from the *University of Siegen* also used to study the hedonic quality of products, and that can be applied to this scenario as well. A manufacturer who builds ugly machines has a bigger sales problem than someone who builds attractive machines. I have the greatest concerns on two levels. First, that people are not only supplemented but replaced for economic reasons, even in such sensitive fields. The other is that manufacturers are taking residents for a ride and opening the door to a kind of fantasy world for them, so that they attribute more to the devices than there is to offer.

**J.L.**  I would say, though, that we shouldn't be prematurely paternalistic there. There have also been studies done with, for example, the robotic seal *Paro*, where residents of care facilities have literally said, "I know that this is a robot, I love him anyway. He's still my friend."

**K.W.**  And I think that's okay. This is, after all, reflective or conscious love, so the person knows what they are doing. I'm concerned about the level that's underneath that, which is unconscious. People might think, out of feelings of loneliness, "at least the robot is still there for me."

**B.H.**  It is somehow inherent in the nature of us humans to form deep relationships with inanimate objects from an early age. Every child does that, having a close relationship with a teddy bear or even a toy car. This behavior somehow belongs to us and will remain even in old age. I find it more difficult when this actually increasingly replaces contact with other people, where it would actually be possible and desirable—just because of economic interest.

**J.L.**  I definitely agree 100%. It is, as in quite a lot of ethical situations, a balancing act. On the one hand, wellbeing and autonomy is definitely increased by the use of such robots. However, there are also dangers that need to be considered, such as the decrease in social interaction and the fear of manipulation. I think that one just has to decide in concrete and individual situations.

**B.H.**  In a human interaction, I also have to deal with frustrations, for example, if my counterpart doesn't fulfill my wishes or doesn't understand me correctly. That is also conceivable with robots. But it could be that the robot adapts perfectly to me and gives me the desired answer, so that my frustration tolerance drops to such an extent that I reduce human interaction more and more and perhaps no longer have any desire for it. The other way around is also conceivable, of course. A robot that always gives me the same answers after a while frustrates me so much that I no longer have any desire for it.

**J.L.**  I think simply that different robots are necessary for different purposes. The good old *Tamagotchi* from the 90s comes to mind for this. If you overfed it or neglected it, it died and could not be revived. This was very frustrating for parents, as they had to spend a lot of money on a new one. One could imagine equivalent projects in the robotics context. A nursing robot that is neglected, for example, could also express its boredom and thus simulate its unavailability. At least as far as it is wanted, depending on individual cases and the needs of the people.

B.H.    If I imagine I'm sitting in a retirement home and my robot starts to feel neglected or shows other indications of dissatisfaction, because it may be programmed that way, then there must also be a way that I can separate myself from this robot.

J.L.    Yes, it's absolutely important, but these are different questions. For these questions of organization, one must perhaps create guidelines. In the case of the robot seal *Paro*, for example, it can evaluate certain manners. If it is stroked, it hums and makes flying movements with its flippers. If, on the other hand, it is shaken, it begins to croak. How *Paro* is handled in a situation can theoretically be stored in its long-term memory. Developing this thought process further, *Paro* could be personalized for each person so that individual needs are taken into account.

B.H.    And it would create a new field of work, for example in psychosocial counseling, in which one might also have to clarify violent relationships between technology and humans. And we have another question about what happens when a resident falls in love and the robot does not return his feelings. We would need a new form of partnership counseling. Here it also comes to questions regarding the responsibility of the manufacturers. If we get into the situation where residents fall in love with a robot, what happens when the robot is no longer supported by the manufacturer and the relationship is thus ended? The robot breaks up with the resident, so to speak. What responsibility does the manufacturer have? What happens then?

B.H.    At first I would say, well, that's like in real life. It's always painful for the person who is abandoned, and they have to process this. On the other hand, with a robot, maybe it could be revived or rebuilt by another manufacturer?

K.W.    An interesting position within robotics says that every robot is unique, even if they all look like they came off the assembly line. That's hard to imagine at first, because the hardware and software is the same. However, each robot is given an individual context in which it is active and learns, builds up individual knowledge, and is thus contextualized. Maybe it gets scratches and bumps, and thus a kind of patina. If that robot is now replaced by an exchange device, these characteristics are gone—the external ones, anyway. The software also cannot necessarily be transferred one-to-one because it is very specific in detail. From a technical point of view, a robot arm doesn't always weigh exactly the same, so the sensor and motor systems always have to be averaged out. This difference is not relevant for the end user. For the technician it makes a big difference, because the software cannot simply be transferred. In this respect, it is unique. Now you might think that this is an enthusiast's idea. But my robotics colleague Antonio Chella from Palermo goes so far as to say that these devices could very well have property rights. After all, we humans protect certain animal species, works of art or monuments, for example. In his opinion, the same should be done for robots. That way, the robot would actually get a kind of identity. If you then go one step further, you end up back at the topic of money, because the use of robots in the care sector, but also in industry, saves money. Of course, they are also productive workers, but so far they don't pay taxes. And since the state also sits hungrily at the table, the identity of the robot will probably be linked to a tax number in the future, so that a tax amount will be due in the end.

J.L. That's what I always notice in these kinds of conversations, that at the end of the day you realize that the questions are actually well known. Whether it's about uniqueness, unavailability and autonomy, or the question of what constitutes a human or humane life. Or also questions about work, i.e., what is decent work and how can we organize it? This made me smile inwardly in the course of the conversation, because in the field of robotics we seem to be facing huge questions that sound almost exotic. But in the end, one ends up with exactly the same questions that we've actually been asking ourselves for several hundred years. I was just thinking in between about a case where people are into balloons. So they are sexually, intimately and emotionally attracted to balloons. This and other cases are unbelievable taboo topics that are bubbling up socially. Because robots are also objects, I feel like these issues are coming back to the surface. But is it a problem that we have relationships with objects? It's actually much more about what is socially considered to be good and acceptable—that's the real problem, not that we're starting to fall in love with our robots.

**PEOPLE ARE FIGHTING TO DISCARD OUTDATED ROLE MODELS. THE ARTIFICIAL INTELLIGENCE OF ROBOTS**

**WORKS WITH DATA FROM THE PAST. ISN'T THERE A RISK THAT THE** ROBOT **WILL REPRODUCE AND REINFORCE OUT-DATED ROLE MODELS?**

K.W.    I would agree; when an artificial neural network is trained, the data comes from somewhere in the past and not from the future. We don't have data, just visions about the future.

J.L.    Yes, absolutely and especially in terms of appearance. My perfect example for this is sex robots. With their specific embodied hardware, clear gender stereotypes reappear. We also find stereotypes in *Siri* and *Alexa*. These stereotypical qualities are all put in there by the company that's developing the technology, often intentionally or even unconsciously. The question is, can we manage to design robots for the care context without such stereotypes, and that are gender neutral? In such a way, more diversity is thus created.

K.W.    I agree. And if you think about the manufacturers, then they are of course also concerned with the fact that they want to work economically, they want to produce something that is suitable for the masses. The things that are suitable for the masses are often conservative, and the majority of that population is mostly older and old-fashioned. They have been socialized differently and have grown up with different role models. From the manufacturer's point of view, an androgynous robot would probably appeal more to younger relatives, while the grandfather in the care bed might prefer a good-looking female care robot.

So it might be worth looking at which groups of people, clusters of people or user segments you're dealing with, and how the requirements differ within these segments. And here I imagine that groups would become visible that could be classified into categories such as age, region, but also intellectual background and socialization. Thereby, wishes relating to the characteristics of the robots could be very different. Here, it could be interesting simply to talk to the caregivers about their experiences.

**IF WE NOW THINK OF PEOPLE FORMING A RELATIONSHIP WITH A SOCIAL** ROBOT **AND MAYBE FALLING IN LOVE—IS THAT MORE OF A UTOPIA OR A DYSTOPIA?**

J.L.    I see it rather as a utopia, in which it does not matter at all which type of being I have an intimate relationship with, whether that's humans, plants or animals—as long as all ethical bases are secured; that is, that they are not minors or being forced against their will. If such exceptions and morally reprehensible things are excluded, I would say, it should be all the same who I love or not. Furthermore, there should be no judgment about whether loving a robot is somehow better or worse than loving a human.

K.W. The answer is that it depends on us and what we make of it. It's a design challenge that needs additional work, using participatory approaches to explore and experiment with how it might be possible to improve quality of life with robots.

In our case of people in need of care, we should openly and pragmatically explore what is attractive and in which area. I see one of the biggest challenges for manufacturers in this open-mindedness. It involves overcoming preconceptions about the world. It starts with education; we must ensure that more reflective competence enters computer science and engineering. To do this, we have to start at the grassroots level. It is not enough to put an ethicist between the engineers for an evening to explain the world to them. We should focus on the self-enlightenment of the technical disciplines and, on the other hand, ensure that reflective sciences come into contact with real design scenarios, and challenge and encourage them there. Then what belongs together will grow together.

B.H. I think that we no longer have the choice to decide. It is no longer a question of whether it would be a desirable or bad future. It's just there and we have to shape the future with it. It is more interesting to ask how we can manage this. There are different approaches to the programming of emotional robots, as you can see internationally. This means that we will always have a situation where different values and norms are programmed into the machines, where different concepts of social relationships and desires collide, just as we experience as humans in contact with people from other cultures. So, I would also support what you have already said, Karsten: that it is important to intensify this ethical discussion with those who develop the devices, so that they realize which world view they are unknowingly implanting in such a robot. It is essential that they know that these are their values and ethical ideas, from which they act, but that they are not necessarily those of other people, groups or cultures. It's important to find ways to enter that exchange and evaluate how successful the negotiation is about what the machine can do.

FALLING IN LOVE WITH A MACHINE

BRIGITTA HABERLAND, KARSTEN WENDLAND, JANINA LOH

# NIKA

### AIM OF OUR RESEARCH PROJECT

The aim of our research is the creation and invention of common interaction patterns for robotic systems supporting elderly people.

### CONTEXT, ROLE AND TASK OF OUR ROBOTS

As examples, we work with three different types of already existing robotic systems (humanoid, zoomorphic, and thing-like). We develop these robots further to be *companions*, to support the interdependence of elderly people, and to stimulate them physically and mentally by playing games and quizzes, for example. Also, our robots work in teams with the nurses.

### WHAT MEANINGFUL HUMAN-ROBOTS INTERACTION MEANS TO US

Users should be aware at all times that it is only a robotic system with which they are interacting. Therefore, it is very important to educate the users, while also making sure they are treated as individuals. We consider biography work and profiles of users to be important.

### WHO WE ARE

Wohlfahrtswerk für Baden-Württemberg; Fraunhofer Institute for Industrial Engineering IAO; University of Stuttgart, Institute of Human Factors and Technology Management, IAT; University of Tübingen, International Center for Ethics in the Sciences and Humanities, IZEW; C&S Computer and Software GmbH.

# I am Listening to You!—How to Make Different Robotic Species Speak the Same <u>Language</u>

*Interview with*

Daniel Ziegler (D.Z.)
Kathrin Pollmann (K.P.)
Sabine Schacht (S.S.)

*By*

Judith Dörrenbächer
Anne Wierling

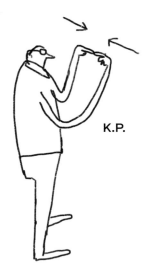

K.P.

**YOU HAVE DEVELOPED SO-CALLED "INTERACTION PATTERNS" IN YOUR PROJECT. WHAT ARE THOSE EXACTLY?**

Interaction Patterns are recurring behavior patterns a robot shows during interaction with humans. We looked at interaction scenarios of different robots, compared them with each other and found that certain actions are repeated. For example, different robots in very different scenarios communicate that they are listening, paying attention, or want to show something. For each of these recurring actions, we created a pattern. For us, the patterns are primarily a stimulus for interaction designers and developers. They are the smallest building blocks that can be used to puzzle together interactions for new robots. We have defined about forty patterns and then summarized them in our pattern wiki (https://pattern-wiki.iao.fraunhofer.de).

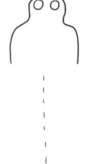

**THE** ROBOTS **YOU LOOKED AT ARE VERY DIFFERENT. ONE IS HUMANOID (PEPPER), ANOTHER ONE LOOKS LIKE A CROSSBREED BETWEEN A DOG AND A RABBIT—IT HAS A ZOOMORPHIC SHAPE (MIRO), AND THE THIRD, THE VACUUM CLEANER** ROBOT **(ROOMBA), DOESN'T RESEMBLE ANY LIVING BEING AT ALL. WHY DID YOU CHOOSE THESE THREE** ROBOTS **AS THE BASIS FOR THE PATTERNS?**

D.Z.   We want our patterns to be transferable—regardless of the form, the interaction ability, or the concrete mode of expression of a robot. We want users to recognize the behavior of different robots, despite the differences in their designs. For example, not all robots have a voice output or a speaker loud enough for every context. Humanoid robots have arms and can gesture, while vacuum cleaner robots, like *Roomba,* on the other hand, lack limbs to express themselves. The challenge was to find a common denominator despite these distinct differences. We had to define a level of abstraction from which the patterns could still be applicable for robot designers.

K.P.   We solved the balancing act between "this should be abstract and transferable but also concretely applicable" by looking at possible communication modalities. For each pattern, we analyzed existing possibilities in human-human, human-animal, and human-technology interaction, or even in robotics. Which best practices and scientific findings already exist? What modalities can a robot use to communicate with humans and express that it is listening? There are many ways for this to be shown, like by means of movement, light, gestures, facial expressions, or sound. Our three robots only served as examples to show how a concrete pattern could be implemented in different robots.

**DO YOU HAVE AN EXAMPLE OF A PATTERN THAT WORKED PARTICULARLY WELL?**

D.Z.   Yes, the pattern *listening* works extremely well for *MiRo* because it has movable ears that it can move back and forth.

K.P.   The patterns also indicate whether a robot's expressive capabilities are sufficient to communicate socially. A *listening* robot means that something is being recorded and this robotic action should be signaled to users. Since *Roomba* has only a few ways to express itself, we have broadened this by adding a ring of LEDs, which was inspired by the luminous ring of *Alexa (Amazon Echo)*. With it, even the thing-like

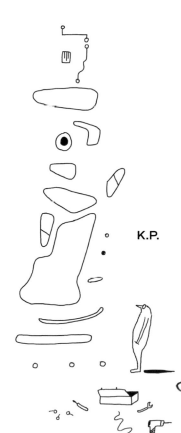

*Roomba* can express that it is listening. An alternative to this solution could be a microphone icon on the display of a thing-like robot that turns toward the direction of the voice being recorded. By always listing the different expression possibilities of a pattern in our pattern wiki, developers can see if possibilities of expression are missing in their robot. This also helps to avoid ethical or legal problems. For example, it prevents the impression that robots are secretly spying on their users.

**THAT MEANS THE PROJECT NIKA WASN'T ABOUT FINDING THE PERFECT INTERACTION SOLUTION FOR A SINGLE ROBOT, SUCH AS PEPPER, BUT ABOUT INVESTIGATING WHAT POSSIBILITIES DIFFERENT ROBOTS OFFER AND HOW THESE CAN BE TRANSFERRED TO OTHER DEVICES?**

**D.Z.**   Exactly. A pattern for *Pepper*, for example, could transfer well to the robot *Nao*, which is also humanoid, even though they don't have the same technical platforms. Nevertheless, not all humanoid robots have the same expressive capabilities, and some of them differ greatly in their abilities. Conversely, this means that the universally formulated and multimodally interpretable patterns must always be translated to a concrete robot. On the one hand, the goal of the patterns is to provide robot developers with a guideline to show how good interaction looks and how it can be designed comprehensibly. On the other hand, the patterns should help robot designers to decide which capabilities the robot should have available and whether further ones could be added with little effort. If designers think about the various capabilities of a robot at an early stage in the design process, it is possible to design the interaction options in a targeted manner.

**K.P.**   In addition, the more a specific robot is marketed commercially, the more patterns it comes with. With *Pepper,* an already quite popular robot, the ears start to light up in a certain way when someone says something. When trying to map this with our own patterns, the expressions often didn't match anymore. Consistent patterns are often difficult to implement because the programming of existing robots often cannot be changed. For truly consistent robot *families* whose robots speak a transferable language, the robots would probably need to be rebuilt completely.

**YOU ALSO WORKED ON PRIVACY ISSUES IN NIKA. HERE, YOU PLAN TO IMPLEMENT A CASCADE MODEL WHERE USERS**

HAVE A CHOICE BETWEEN DIFFERENT SETUPS WITH DIFFERENT LEVELS OF FUNCTIONALITY. THE BASIC FUNCTION WOULD MEAN THE FEWEST FUNCTIONALITIES AND NO PERSONALIZATION, BUT CONSEQUENTLY THE MOST DATA-EFFICIENT PROCESSING OF PERSONAL DATA. LET'S ASSUME A USER WITH A ROBOT HAS A VISITOR. BECAUSE THERE IS LESS OF AN ISSUE OF CONSENT IN THE CONTEXT OF DATA PROTECTION FOR THE VISITOR IN THIS CASE, DOES APPLYING YOUR CASCADE MODEL BRING ADVANTAGES TO THE USER?

S.S. Yes, because the cascade model means that the basic function can be used if necessary. In addition to the user, there could always be third parties in the room, such as family, friends or caregivers, who have not consented to data processing. Their personal data would then only occur as a by-product and would not be evaluated, which would also make it conceivable that another permissible circumstance, such as the balancing of interests pursuant to Art. 6 (I) f DSGVO, would apply to the processing.

 WHAT EFFECT DOES THE DEFINITION OF THE CASCADE HAVE ON THE PATTERNS, AND CAN THE REQUIREMENTS OF THE GDPR REGARDING DATA PROTECTION-FRIENDLY DEFAULT SETTINGS PURSUANT TO ART. 25 BE REALIZED BY MEANS OF THE VARIOUS CASCADES?

D.Z. If a certain functionality is not permitted in the corresponding cascade, such as the use of the cameras, then the robot— regardless of whether it is *Pepper* or *Roomba*—can neither recognize people nor turn towards them. That is, the robot would then have to exhibit an alternative behavior to express attention. Based on the patterns, other modalities can be identified that have the same goal. For example, if a user is playing a quiz game with the humanoid robot *Pepper* and they do not allow recording with the microphone, then the answers of the quiz could also be displayed or entered on *Pepper*'s tablet. That would not be optimal, but it is an alternative.

K.P. If certain functionalities are not available for data protection reasons, then the robot's range of functions are restricted. Some scenarios cannot be realized if neither the camera nor the microphone can be switched on. Each specific use case requires analysis for what the interaction flow is and whether the function can be replaced in a meaningful way, so that it conforms to what users allow.

**THE GDPR REQUIREMENTS ARE FUNDAMENTALLY CHAL-LENGING FOR ROBOTICISTS. DO YOUR PATTERNS OFFER ADVANTAGES IN DEALING WITH THE REGULATIONS?**

D.Z.     The patterns can be used as an analytical tool to clarify when the robot needs the microphone, for example. In this way, a structure can be created that describes the purpose and scope of the processing. This makes it easier to comply with the requirement in Article 30 of the GDPR, for example, to create a list of processing activities. Patterns also mean that users no longer have to make classic "all-or-nothing" decisions, and this has a beneficial effect on compliance with voluntariness. Decisions become much more fine-grained in the interaction. Ideally, the robot offers alternative solutions instead of completely eliminating the function. For example, the robot might say, "hey, if I get to use my cameras now, then I can offer to automatically identify who you are in this situation." This conveys the purpose more intuitively than a page of text which hypothetically describes the situation before the robot even exists. This leads to a lower cognitive load for users when providing information as part of the consent process.

Inspired by the interview, the illustrator Johanna Benz visually commented on chances, risks and scenarios of robotics from her own perspective. © Johanna Benz

# How to Really Get in Touch with Robots—Haptic Interaction Technologies for VR and Teleoperation

Bernhard Weber
Thomas Hulin
Lisa Schiffer

Imagine if you were interacting with a virtual robot and you could actually feel the robot when shaking hands in the virtual scene. Today, haptic interaction technologies make it possible to create an experience of touch by, for example, applying forces or vibration to the user. This applies to interactions with robots that are simulated in a virtual scene, but also to real robots that are teleoperated by a human (i.e., → telerobotics). In both cases, the human operator directly controls a virtual or real robot using a haptic device. Such an interaction device allows the robot to be controlled while at the same time interaction forces between the robot and its environment are fed back to the user in the form of forces or vibrations. Telerobotic systems are used in environments that are too dangerous for humans (e.g., disaster response, bomb disposal), or in environments that are difficult to access (e.g., keyhole surgery, →Fig. 1).

**Telerobotics** means a human operator controls the movements of a robotic system (e.g., rover, humanoid, robotic arms) from a distance.

In telesurgery, a surgeon operates a robotic system on the operating table, moving the surgical instruments and endoscope inside the patient's body by means of robotic arms. The surgeon is usually in the same operating room, but not directly at the operating table.

HOW TO REALLY GET IN TOUCH WITH ROBOTS

BERNHARD WEBER, THOMAS HULIN, LISA SCHIFFER

The concept of telesurgery originated in military scenarios where surgeons needed to be able to operate on wounded soldiers from a safe distance. In civilian use, however, equally interesting application scenarios open up, such as a telesurgical intervention by an expert on a patient in another hospital, on another continent, or even on a space station.

In telesurgery, it is of course also very important that the interaction between the robot and its environment is tangible. For example, the surgeon needs to be able to feel the relevant forces during cutting, or when pulling the thread during suturing or knotting. When force is applied to the tips of the surgical instruments, this force is measured and transmitted to the force feedback devices which generate a resistive force. Not surprisingly, many empirical studies comparing virtual reality (VR) and telerobotic systems with and without force feedback show that human operators show improved performance when force feedback is present, because they are able to adjust the interaction forces more accurately than they can when solely acting on visual information (Weber and Eichberger, 2015).

# MAY THE FORCE BE WITH YOU— SPACE ROBOTICS AND TERRESTRIAL APPLICATIONS

Sometimes it is danger rather than an inaccessible environment that makes the use of teleoperation technology advisable. On space missions, astronauts face extreme hazards such as cosmic radiation. In the future, it is expected that astronauts will explore the surface of the Moon, and later Mars, ultimately establishing habitats there. Robotic systems will also be used on these missions, and these will be controlled from orbital spacecraft by astronauts. Although some robots will be able to operate autonomously (i.e., without any human control), teleoperation will still be necessary for some such tasks in the future. In these cases, astronauts, like surgeons, will depend on adequate and realistic force feedback to solve telerobotic tasks efficiently and safely. A sensory teleoperation experiment was conducted by the German Aerospace Center (DLR) in 2015 as part of the *Kontur-2* project, in which several robotic systems (on the test planet Earth) were successfully teleoperated by Russian cosmonauts using force feedback joysticks from the *International Space Station (ISS)* (→Figs. 2, 3, and 4, Riecke et al., 2016).

Haptic interaction technology also has great potential for terrestrial robotic assistants. Although the vast majority of robotic daily living assistance initiatives and projects involve autonomous systems, there will always be situations in which it makes sense for humans to be able to teleoperate such systems using haptic feedback.

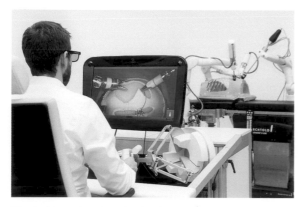

**Fig. 1** Example of a telerobotic system using haptic interaction technology: DLR's telesurgical system MiroSurge® with two force feedback devices (sigma.7 devices, Force Dimension) at the surgeon's console on the left side of the picture, and robotic arms at the operating table on the right. © DLR, CC-BY3.0

**Fig. 2** Cosmonaut controlling robots on test planet Earth with a force feedback joystick from the International Space Station (ISS). © Roscosmos

**Fig. 3** With this joystick, the cosmonaut Sergey Volkov, on board the ISS, was able to feel the force of a telerobotic handshake. © DLR, CC-BY3.0

**Fig. 4** Volkov's wife greeting and shaking hands with her husband on board the ISS. © DLR, CC-BY3.0

HOW TO REALLY GET IN TOUCH WITH ROBOTS

BERNHARD WEBER, THOMAS HULIN, LISA SCHIFFER

**Fig. 5**  DLR's stationary HUG system.
© DLR, CC-BY3.0

**Fig. 6**  The omega.6 (Force Dimension) desktop device.
© Force Dimension, Switzerland

**Fig. 7**  The Falcon (Novint) desktop device.
© DLR, CC-BY3.0

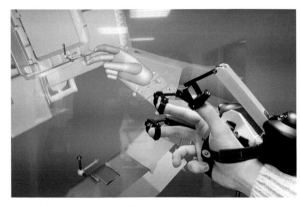

**Fig. 8**  The wearable HGlove. © Haption

Telerobotics increases the potential for social interaction, if one thinks, for example, of elderly people living alone or patients in quarantine wards. Through telerobotics, not only medical staff and caregivers but also relatives can make contact. From a technical perspective, many robotic systems will continue to fail in the future due to the variety and complexity of everyday tasks and environments, and telerobotic solutions would enable humans to help out. In both these scenarios, the incorporation of haptic interaction technology will be very important, allowing the teleoperating human to enter into a physical interaction and directly perceive the interaction forces between the robot and its environment. For example, an operator could safely help an elderly person stand up, feeling exactly what forces were occurring. Also, in complex and confusing situations where, for example, transmitted video information is insufficient, additional haptic feedback can be extremely helpful. A robot's field of view is often restricted; lighting conditions in a remote environment may be suboptimal, or the bandwidth of a video channel may be limited. For these reasons, it is all the more important for teleoperating humans to have access to haptic information.

# FORCES AND/OR TACTILE FEEDBACK?

Having illustrated potential use scenarios for haptic interaction technology, a further question involves which haptic devices are best suited for which tasks. Two main types of system solution exist for representing haptics in VR: systems that can output forces, i.e., the device is capable of generating push or resistance (these include force feedback joysticks); and systems that can provide tactile feedback such as vibration. Let us focus on force feedback devices first. The advantage of these devices is that relatively high levels of force can be generated, creating the impression of real contact during interactions in VR. Currently, several devices are commercially available; Wang et al. (2019) provide a comprehensive overview of force feedback devices for VR applications. These can generally be categorized into systems which are set up as workstations (stationary systems, e.g. →Fig. 5), smaller systems which can be placed on a desk (desktop devices, e.g., →Figs. 6 and 7) and devices which can be worn directly on the body (wearable devices, →Fig. 8). Due to their large workspace, stationary systems are well suited to controlling large systems like humanoid robots. For applications focusing on fine motor control, as in telesurgery, desktop systems are more appropriate. For the most natural type of interaction, with the least restriction from the input device, wearable systems (such as *Haption's HGlove*, which provides force feedback to three fingers) are an interesting alternative.

Let us take a closer look at tactile feedback systems. These are an interesting alternative to force feedback systems, as they are available at a much lower cost, and are already widely used in the consumer market, e.g., in game consoles, whereas force feedback devices can easily cost several thousand euros and are therefore almost exclusively used in professional applications where the benefits outweigh the costs. Tactile devices are not only cost-effective, but they are also usually lighter, more compact and therefore easily portable. A particular strength of tactile systems is their combinability. For example, several wristbands can be combined to output tactile stimuli on both arms, or a wristband can be combined with a force feedback device. Since such force feedback devices are often guided by hand, the forces are of course only perceived there. Collisions, e.g., with the elbow in a virtual scene, can then be displayed in a vibrotactile way, so that whenever the human operator's elbow touches a surface or object in the virtual scene, a vibration is generated by the wristband's motors, thus warning the operator (Sagardia et al., 2015) (→Figs. 9 and 10). In telesurgery, for example, manual interaction forces should be represented as realistically as possible, i.e., preferably as force feedback. Forces acting on the robot arm itself (e.g., collisions of the robot arm with the environment) could be represented in a vibrotactile manner.

Although they are not able to produce real forces, vibrotactile feedback systems nevertheless improve performance in basic experimental tasks (Kontarinis and Howe, 1995; Massimino, 1992), even though the positive effect is weaker compared to force feedback (Nitsch and Färber, 2012, Weber and Eichberger, 2015). While vibrotactile feedback is an effective tool for displaying simple haptic information (e.g., when there is a collision), it is still only a substitution for real interaction forces. For applications which do not require high-resolution force feedback, but where the user should simply be warned, e.g., that s/he is colliding with a virtual obstacle or is entering a sensitive area, vibrotactile feedback is a sufficient and reasonable alternative to force feedback.

# EVALUATING HAPTIC INTERACTION TECHNOLOGY

Whether, and to what extent, a specific haptic interaction device is effective for a given use case is evaluated through empirical studies, measuring human performance as well as subjective elements such as perceived workload, sense of presence, and usability in experimental user studies. At DLR, for example, an evaluation study was conducted comparing visual, vibrotactile and force feedback for collisions in VR. Participants had to perform a virtual assembly task

**Fig. 9** Combined use in a virtual assembly task of the force feedback device HUG with the vibrotactile device VibroTac (Schätzle et al., 2010) circled in red. © DLR, CC-BY3.0

**Fig. 10** A close-up view of the VibroTac. © DLR, CC-BY3.0

**Fig. 11** VR simulation of robotic arms/hands and a satellite. © DLR-SC, CC-BY3.0

**Fig. 12** Control station, including the force feedback device HUG and a head mounted display with head tracking functionality. © DLR, CC-BY3.0

BERNHARD WEBER, THOMAS HULIN, LISA SCHIFFER

including, for example, the insertion of a virtual plug into a socket. The results showed that the high-resolution force feedback provided by the *HUG* system was most beneficial in terms of precision, workload, and spatial orientation compared to the other feedback modalities (Weber et al., 2013). In a similar VR study, DLR compared different commercially available and prototypical force feedback devices (Schneider, 2015), also finding that DLR's *HUG* system was superior in terms of perceived authenticity of contact due to the comparably high stiffnesses the system is able to produce. The ease of operation of desktop devices like *Force Dimension's omega.6 or sigma.7* and the *Novint Falcon* device, however, allow for better fine manipulation performance and also reduce workload. In the above-mentioned space project *Kontur-2*, a simulation environment was also used to investigate whether, for example, human operators were still able to perform basic motion tasks with a force feedback joystick during spaceflight (Weber et al., 2021).

# EVALUATION AND TRAINING OF HAPTIC INTERACTION IN VIRTUAL ENVIRONMENTS

As argued above, it is particularly useful in the context of research and development to evaluate haptic interaction concepts in VR. VR simulations are an appropriate training tool for fully developed robotic systems, and are often used for this purpose (e.g., in telesurgery: Bric et al., 2016). In recent years, for example, DLR has exploited the potential of VR as a training tool for applications in space robotics. It is well known that space debris within Earth's orbit (e.g., derelict satellites and spacecraft, fragmentation debris) is an increasing problem in spaceflight. Therefore, there is an urgent need for maintenance, repair, and de-orbiting technologies. DLR has developed a virtual reality framework which allows robotic maintenance tasks on satellites to be simulated interactively (Sagardia et al., 2015). In addition to a realistic VR simulation (→Fig. 11), an immersive interaction concept was implemented which included force feedback technologies, i.e., collisions in the VR were conveyed at the user interface. A special feature of this VR simulator was the implementation of a high-performance algorithm developed at DLR to calculate the collision forces in the virtual scene in real time (i.e., *Voxmap PointShell* algorithm: McNeely et al., 2005; Sagardia, 2019), allowing very strong, realistic contact forces (e.g., collisions between the robot and the satellite's surface in VR) to be simulated without any noticeable delay. Here, DLR's bimanual haptic device *HUG* is used as an interaction device. The *HUG* system consists of two robotic arms which are moved by the human operator to control a remote, or in this case virtual, robotic system (→Fig. 12).

There are several clear benefits to using DLR's robotic arms. First, they are—compared to industrial robots—lightweight and thus easy to move. Second, a large workspace is available and the user is able to interact without notable restrictions. Third, the system is not only able to provide realistic haptic feedback to both human hands, but forces applied by the human operator are also precisely detected by sensors integrated into each of the robot's joints. With an additional head-mounted display, the user experiences a strong feeling of presence in the virtual scene.

This is facilitated by the fact that the motions of the user's head are tracked by cameras and these motions are transferred to the virtual scene, i.e., the field of vision is naturally adapted. DLR's VR simulator can be used to verify whether and how certain maintenance tasks on satellites, space stations or planetary habitats can best be performed telerobotically, and to train future astronauts or maintenance engineers.

# ASSISTIVE ROBOTICS WITH HUMAN TOUCH

The future of assistive robotics is often painted as a scenario in which autonomous robotic systems support individual everyday tasks. Teleoperation introduces an alternative dimension to human-robot interaction, and represents an important complement to purely autonomous operation. Integration of the haptic sense enables humans to experience a feeling of →telepresence, i.e., they have the realistic impression of slipping directly into the robot's skin. This makes it even easier for humans to interact naturally with the remote environment, experience real contact with others at a remote site, and complete various types of tasks more effectively. VR offers a perfect platform for training, evaluating and generally exploring this form of HRI.

**Telepresence** means that the human operator feels physically present at the remote robotic site. This feeling can be created if sufficient information from the robot and its environment is available and displayed in a sufficiently natural way.

Bric, J. D., Lumbard, D. C., Frelich, M. J., & Gould, J. C. (2016). Current state of virtual reality simulation in robotic surgery training: A review. *Surgical Endoscopy, 30*(6), 2169–2178.

Kontarinis, D., & Howe, R. (1995). Tactile display of high-frequency information in teleoperation and virtual environments. *Presence, 4*(4), 387–402.

Massimino, M. (1992). *Sensory substitution for force feedback in space teleoperation* [PhD thesis]. MIT, Department of Mechanical Engineering.

McNeely, W. A., Puterbaugh, K. D., & Troy, J. J. (2005). Six degree-of-freedom haptic rendering using voxel sampling. *ACM SIGGRAPH 2005 Courses*, 42–es.

Nitsch, V., & Färber, B. (2012). A meta-analysis of the effects of haptic interfaces on task performance with teleoperation systems. *IEEE Transactions on Haptics, 6*(4), 387–398.

Riecke, C., Artigas, J. Balachandran, R., Bayer, R., Beyer, A., Brunner, B., Buchner, H., Gumpert, T., Gruber, R., Hacker, F., Landzettel, K., Plank, G., Schätzle, S., Sedlmayr, H.-J., Seitz, N., Steinmetz, B.-M., Stelzer, M., Vogel, J., Weber, B., Willberg, B., & Albu-Schäffer, A. (2016). *KONTUR-2 Mission: The DLR force feedback joystick for space telemanipulation from the ISS* [paper presentation]. International Symposium on Artificial Intelligence, Robotics and Automation in Space (i-SAIRAS 2016), Beijing, China.

Sagardia, M., Hertkorn, K., Hulin, T., Schätzle, S., Wolff, R., Hummel, J., Dodiya, J., & Gerndt, A. (2015). VR-OOS: The DLR's virtual reality simulator for telerobotic on-orbit servicing with haptic feedback. *2015 IEEE Aerospace Conference* (pp. 1–17).

Sagardia, M. (2019). *Virtual manipulations with force feedback in complex interaction scenarios* [Doctoral dissertation]. Technical University of Munich.

Schätzle, S., Ende, T., Wüsthoff, T., & Preusche, C. (2010). VibroTac: An ergonomic and versatile usable vibrotactile feedback device. *Proceedings of 19th International Symposium in Robot and Human Interactive Communication*, 705–710.

Schneider, A. (2015). *Evaluation of haptic human-machine interfaces for virtual reality applications* [Master's thesis]. Technical University of Munich.

Wang, D., Ohnishi, K., & Xu, W. (2019). Multimodal haptic display for virtual reality: A survey. *IEEE Transactions on Industrial Electronics, 67*(1), 610–623.

Weber, B., Sagardia, M., Hulin, T., & Preusche, C. (2013). *Visual, vibrotactile and force feedback of collisions in virtual environments: Effects on performance, mental workload and spatial orientation*. In Shumaker, R. (Ed.), Virtual, Augmented and Mixed Reality /HCII 2013, Part I, LNCS 8021 (pp. 241–250). Springer.

Weber, B., & Eichberger, C. (2015). The benefits of haptic feedback in telesurgery and other teleoperation systems: A meta-analysis. In Antona, M., and Stephanidis, C. (Eds.), *Universal Access in Human-Computer Interaction. Access to Learning, Health and Well-Being. Part III, LNCS 9177* (pp. 394–405). Springer.

Weber, B., Riecke, C., & Stulp, F. (2021). Sensorimotor impairment and haptic support in microgravity. *Experimental Brain Research, 239*(3), 967–981.

# Part 2
# Designing Future Environments—

# Social
# Innovation
# Initiated
# by Robots

# Design Fiction—The Future of Robots Needs Imagination

Ronda Ringfort-Felner
Robin Neuhaus
Judith Dörrenbächer
Marc Hassenzahl

*Early on a Wednesday morning in 2045. The mall fills with the first customers and robots. Businesspeople pick up their pre-ordered breakfast from robots. Families settle down in the café, enjoying fresh coffee and croissants, while robots run their errands or play with their kids. A jogger is doing his laps in the rooftop garden, while robots do his weekly grocery shopping. Teenagers hang out in the arcade and giggle at the clothing suggestions sent to them by a robot. The mall offers stores to buy everything imaginable, as well as a variety of opportunities for leisure and recreation. Robots are ubiquitous—high-tech assistants, whose sole purpose is to make the shopping experience as pleasant as possible. Do you notice how the robots get to know the visitors, their clothing size, their likes and dislikes better with each visit? How they flawlessly adapt their behavior to their acquired knowledge? Recently, many of the premium shopping malls have started to cooperate and exchange the data collected by their robots.*

This chapter is about how design can be used to speculate, elaborate, and evaluate potential futures with robots. It is about how we can use design to open up new perspectives, to create spaces for debate, to inspire people's imaginations, to thoroughly explore different futures—the desirable and the undesirable. In this context, the scenario described above is not just a fictional story written to entertain, but the result of a design practice called →Design Fiction. Design Fictions are created by designers and researchers to raise questions about

DES GN FICTION

RONDA RINGFORT-FELNER, ROBIN NEUHAUS, JUDITH DÖRRENBÄCHER, MARC HASSENZAHL

**Design Fiction** is "part of a group of design practices [called] Speculative Design. Rather than solving problems, these approaches use design to ask questions. They do this by creating prototypes, but instead of making prototypes that will later be put into production, these prototypes are used to encourage people to think critically about issues that the design embodies" (Coulton et al., 2019, p. 9).

**Possible Futures** include preferable futures, but also probable or plausible futures. A possible future makes a link between today's world and the suggested future world. "A believable series of events that led to the new situation is necessary, even if entirely fictional. This allows viewers to relate the scenario to their own world and to use it as an aid for critical reflection" (Dunne & Raby, 2013, p. 4).

how everyday life with certain technologies will and should be, highlighting potential problems and opportunities—long before the technology is mature enough to become a part of our lives.

Certainly, many filmmakers and novelists have already thought about the future with robots. Almost everyone is familiar with various types of robots from novels or movies; mostly they rebel against humans or against being mistreated by humanity, as in *Blade Runner* (Scott, 1982), *I, Robot* (Proyas, 2004) or the popular television series *Westworld* (Nolan & Joy, 2016). Of course, these stories and images shape our perception of what a future with robots might look and feel like. Unfortunately, most of these cinematic depictions are far removed from what is desirable, or what we can expect soon. Hopefully, we will never have to deal with a robot pointing a gun at us in our living room. On the contrary, roboticists are currently working on a technological species which will live with us in our homes like family members (→ p. 78), go shopping with or for us (→ p. 154), work side-by-side with caregivers in nursing homes (→ p. 216), or support therapists in autism therapy (→ p. 58). So far, these robots are most often found in research laboratories, at tech fairs or in highly supervised study settings. But what will happen when they are deployed into the real world? Will we accept them into our families? Will they increase sales, make our lives easier, and positively impact our wellbeing? Or will they lead to lower sales, more conflict, and even social isolation? And are there any ways to mitigate such negative impacts?

While roboticists think primarily about a robot's design, its exact configuration and technical implementation, far-reaching social consequences tend to be slightly out of focus. Other researchers, such as ethicists, psychologists or sociologists, discuss the moral and social questions surrounding that technology, but they often do so in a rather general way, far removed from concrete designs and a design's impact on everyday life.

With this in mind, in April 2021, I (first author), together with Jochen Feitsch from the *University of Applied Sciences of Düsseldorf*, organized a robot Design Fiction workshop, where we learned how valuable early speculation about future mundane life with robots can be for robot developers, as well as for laypeople who might live alongside such robots in the future. Together with developers of four robots (*VIVA, I-RobEka, ERIK, KoBo34*), we created four → possible futures based on their ideas of how their robots might be designed and used in the year 2045. We created various Design Fiction artifacts—texts about future scenarios, like the one at the beginning of this chapter, and fictitious product flyers (→ Figs. 1 and 2)—which assume that the imagined technology is already widely used.

We created these artifacts in advance, together with designers, copywriters, and the robot developers, to make the potential futures more graspable.

We then invited the robot developers, as well as researchers and role-players, to our robot *Design Fiction workshop* to speculate on and discuss these possible futures and how we might live in them. During the workshop, participants used the artifacts to immerse themselves in the four different futures and their respective robots, and then discussed their thoughts. By reading through the flyers and future scenarios, a concrete image of each future was gradually formed, becoming more and more detailed through the participants' thoughts and imaginations. Subsequently, the participants were asked to imagine different people whose lives might be affected by the robot in the future. They imagined a wide variety of individuals, ranging from a child, mother, janitor, health insurance representative, tech guru, politician, candy maker, pet, and even the robot itself. The participants were asked to put themselves in the shoes of each fictional character to explore likely emotions, attitudes and opinions, and to imagine fictional everyday experiences of these characters.

In a "future-dialogue," participants discussed with each other from their character's perspective. For example, one researcher became a 7-year-old child who sadly recounted how her brother had stopped playing with her because of the family's social robot. Another participant became a therapist who reported on the consequences of social isolation, but also on the potential for better therapy through access to more data. Someone else spoke from the perspective of a tech guru, enthusiastically seeking to bring the emotion recognition module developed for autistic children to the entire population to help humanity progress to a new level. Through this imagination and role-play, diverse topics, conversations, conflicts, and questions were raised: How must robots be designed to promote social interactions rather than restricting them? Does the incorporation of seemingly positive characteristics, such as conflict avoidance, into a robot's design lead to long-term benefits, or do we need robots which will provide criticism and start debates? The process of discussing futures from different points of view revealed areas of tension, far-reaching ethical and social implications, and ultimately concrete solutions (such as training for social interaction with robots)—themes that are rarely thought about in regular development processes, since roboticists focus on the robot itself and not on the possibilities, social dynamics and potential conflicts initiated by their robots.

You might object that the entire discourse, with all the conflicts and issues that emerged, was only fictional. However, the emerging fictions remained strongly grounded in everyday real life. The participants did not simply make things up, but rather *simulated* futures based on their knowledge about their own everyday practices,

DESIGN FICTION

Discover our new version of VIVA, the social robot that offers more harmony and well-being in your everyday life!

Now available in a number of new designs!

Here in the VIVA store on the second floor.

**Fig. 1**  Fictional product flyer showing future vision for the social robot VIVA.
© University of Siegen, Ubiquitous Design & University of Applied Sciences Düsseldorf

**Fig. 2**  Fictional product flyer showing future vision for the shopping robot I-RobEka. © University of Siegen, Ubiquitous Design & University of Applied Sciences Düsseldorf

RONDA RINGFORT-FELNER, ROBIN NEUHAUS,
JUDITH DÖRRENBÄCHER, MARC HASSENZAHL

needs, and knowledge of how technology affects particular situations. This led to fictional, yet plausible themes. Of course, none of this represents a prediction about what precisely we can expect in the future. Rather, it is a pre-enactment, a simulation of the possible feelings, reactions and opinions of relevant stakeholders— but nonetheless, one from which we can learn. By evaluating our pre-enactment, we learned about the positive and negative impacts a robot might have on our lives, and identified aspects that might prevent robots from being accepted. If we are aware of these things, we can use this knowledge to actively shape and consciously direct the future through design.

Of course, this is not easy. After all, the whole process of thinking up possible futures and speculating about them requires fantasy, imagination and the will to detach oneself from preconceived concepts. Critical speculation and Design Fiction plays an important part here, providing a frame to do so.

# DESIGN AS A WAY OF SPECULATING

When thinking about design, many people assume it to be mostly about future commercial successes and solving concrete problems relating to form, interaction, or even packaging. We are used to pictures and movies of commercial products that do not yet exist, such as drones for delivering packages, or 3D printers for printing food. Usually, designers focus on the product itself to be commercialized, rather than the potential consequences of its use, or alternative concepts. Nevertheless, design can also be used to critically speculate about a product's impact, and about ideas not primarily intended for market exploitation, but rather concerned with broader questions of how we want to live in the future.

In a society obsessed with fast-paced innovation and practical, problem-solving products, designers and industry understandably feel the need to create market-oriented products. However, this can ultimately cause us to move in circles, as alternative ideas may not seem *practical* enough, or potentially commercially successful enough, for designers to spend time exploring them. We need to carve out a space to explore these ideas to challenge existing expectations about the future, and Speculative Design and Design Fiction are ideal for this.

In fact, → Speculative Design is a more critically minded design practice that is based on fiction to explore, discuss, and reflect on ideas beyond the market-focused or obviously practical. Speculative Design thrives on imagination and focuses on asking questions rather than providing answers (Dunne and Raby, 2013).

**Speculative Design** is a general term of a design practice that comprises a range of similar design practices such as Critical Design or DesignFiction. "Speculative Design practices have no direct interest in producing a finished article for production, sale or implementation. [They] aim to challenge assumptions, be critical, and stimulate conversations. Design Fiction sits within the taxonomy of Speculative Design" (Lindley et al., 2015, p. 59).

Dunne and Raby (2007) provide an excellent example of Speculative Design in the domain of robots. In their *Technological Dreams Series: No. 1, Robots*, they present thoughts about alternative emotional interactions with household robots. They show four different robots which address human fears created by years of negative portrayals of robots in science fiction. For example, concerns and fears about privacy and data protection are addressed by Robot 3 (→ Fig. 3), which uses retinal scanning technology to decide who can access its data. While in movies iris scanning is always based on a quick glance, here, a long look is required; the robot has to be sure it's you. Robot 4 (→ Fig. 4) is a very needy robot which, although very intelligent, is trapped in an underdeveloped body and is therefore highly dependent on its user. This neediness built into a highly intelligent product allows the user to maintain a sense of control, thus addressing a human fear of losing control to robots. This project offers a new perspective on domestic robots, their form and interactions, and critically asks how we might interact with robotic household members in the future. Dunne and Raby write: "These objects are meant to spark a discussion about how we'd like our robots to relate to us: subservient, intimate, dependent, equal?" (2007) At first glance, the robots appear strange, but in a deliberate way. They create friction, an encouragement to consider the motives of the designers rather than asking what the robot can do in terms of its functionality.

Thus, Speculative Design (including Design Fiction) is not about creating innovations, market successes or improving current products, but about critically discussing how things ought to be in the future. Physical prototypes are developed, not for the purpose of putting them into production, but to provoke discussions, and to make these discussions more real by providing concrete forms, images and situations. Speculative Design poses questions but does not claim to provide answers; rather, the answers might be delivered by those who participate in the process. This, along with the absence of an intention to develop actual marketable products may, in fact, be the most criticized aspect of Speculative Design practices—idle ideas, many questions and a vague method which seems to lack any aim to actually solve anything. In the words of James Auger: "My response to this is that the methodology is not based on short-term solutions but has the purpose of providing a longer-term, more critical approach to technological research and development" (2014, p. 21). By exposing design decisions to the complex rules of daily life before a product actually exists, and by asking the right questions when confronting a broad audience with it, we create the opportunity to inform and redirect technological developments.

RONDA RINGFORT-FELNER, ROBIN NEUHAUS, JUDITH DÖRRENBÄCHER, MARC HASSENZAHL

**Fig. 3** Anthony Dunne & Fiona Raby, Technological Dreams Series: No. 1, Robots, 2007. Here: Robot 3: Sentinel. © Per Tingleff

**Fig. 4** Anthony Dunne & Fiona Raby, Technological Dreams Series: No. 1, Robots, 2007. Here: Robot 4: Needy One. © Per Tingleff

**Fig. 5** Excerpts from the fictional documentary about a social robot Seyno (2018). Here: A developer explains the (fictional) technical background of Seyno. © Ronda Ringfort-Felner, Valentin Puls

**Fig. 6** Excerpts from the fictional documentary about a social robot Seyno (2018). Here: A traveler shares his (fictional) experiences with Seyno. © Ronda Ringfort-Felner, Valentin Puls

RONDA RINGFORT-FELNER, ROBIN NEUHAUS,
JUDITH DÖRRENBÄCHER, MARC HASSENZAHL

# SPECULATING ABOUT SOCIAL ROBOTS

Think of potential social robots. It is striking that while companies try to sell them, they often lack a clear notion of their specific *meaning-fulness*. Some, such as *Sony's Aibo*, just mimic pets. But beyond the fact that real pets cannot be switched off, the additional contribution of an *Aibo* to everyday life remains vague. It seems as if the interest in the technology itself drives purchase (if at all).

In the seminar *Critical Inquiries into Human-Robot-Interaction*, graduate students of Human-Computer Interaction at the *University of Siegen* were asked to speculate about potential roles and functions of future social consumer robots. For example, together with Valentin Puls, I (first author) created a concept for a social robot called *Seyno* (Ringfort-Felner and Puls, 2018). It is supposedly used on high-speed trains by the national railway company of Germany and is tasked with reminding passengers to comply with social and societal norms. *Seyno* confronts passengers with their inappropriate behavior, such as listening to music that is too loud, eating particularly smelly food, or putting their feet up on seats, when other travelers might feel incapable or afraid of addressing such issues. The idea to create a robot for this particular use may seem strange at first, but unlike humans, robots are emotionally invulnerable, have no sense of shame, and are not affected by any emotions when communicating. For a human, such tasks would require rather *supernatural social powers*, but this is not the case for a robot. Valentin Puls and I explored this idea through a fictional documentary (https://vimeo.com/291653885), which features different stakeholders, such as a CEO, railroad manager, and various travelers (→ Figs. 5 and 6). They talk about their imagined experiences with and feelings towards the robot. The robot is not shown in the film (an example of → Invisible Design). Instead, we focus on the conceptual idea and how it might impact travel. The work creates a concrete vision of the potential impact of a *meaningful*—in terms of the social, not necessarily in terms of the commercial—robot, without the need to deal with its form and interaction design. At the same time, through the fictional experiences and opinions of the travelers, the work critically examines the extent to which a robot can, or should, actually take over socially unpleasant tasks for humans.

**Invisible Design** is the mediation of user experience without explicitly depicting the object that generates that experience (Briggs et al., 2009).

Although exploration of alternative ideas and their implications through Speculative Design and Design Fiction seems valuable, the practice is rarely used in robotics development. While it is often difficult to see how existing social robots make a clear contribution to society, which is why they are not particularly successful, unique, and new ways of using robots often seem strange and superfluous. In the seminar *Critical Inquiry into Human-Robot Interaction*, a number of other interesting concepts emerged, such as a couch potato robot sponsored by *Netflix* to make movie watching more social, a robot to support immigrants with acquiring tacit German cultural knowledge, or a robot to help couples who argue a lot. Of course, many of these ideas are questionable. However, they represent valid application scenarios for social robots, making use of the unique possibilities robots offer.

# DESIGN FICTION IN RESEARCH AND PRACTICE

While Speculative Design comprises, or is related to, a number of similar practices, Design Fiction is one that emphasizes a "character of neutrality (as opposed to criticality)" (Lindley et al., 2015, p. 59). Design Fiction possesses a narrative element (Kozubaev et al., 2020) and creates emerging or fictional technologies which are brought into the real world through artifacts, such as videos, posters or quick start guides (Coulton et al., 2017). It puts people at the center (Lindley et al., 2015) and depicts futures where a particular technology is already widely used to help develop an understanding of its potentially far-reaching ethical and societal implications (Coulton et al., 2019; Lindley et al., 2017).

A well-known example of Design Fiction comes from the *Near Future Laboratory*, a team of designers and researchers. They created a fictional *IKEA* catalog that features various imaginary Internet of Things (IoT) products (→ Fig. 7), such as a smart sofa that adjusts its color and pattern depending on the mood of the person sitting on it (→ Fig. 8), dynamic cooking instructions on a countertop (→ Fig. 9), or a shampoo that autonomously re-grows through eco-friendly biogenerating base starter compounds (Near Future Laboratory, 2015). Like all Design Fictions, the catalog is meant to "encourage conversations about the kinds of near futures we'd prefer, even if that requires us to represent near futures we fear" (Near Future Laboratory, 2015). The catalog was deliberately chosen as a compelling way to represent normal, ordinary life in many parts of the world. It combines the practical production of a printed catalog with a variety of microscale scientific, technological, and social fictions expressed through product descriptions, asides, disclaimers or footnotes. Reading through this familiar type of catalog blurs the borders between reality and imagination.

Hearing the term Design Fiction for the first time, you may have thought immediately of → science fiction. In fact, the term was coined by the science fiction author Bruce Sterling, who once mused that he doesn't write *science* fiction, but *Design* Fiction (Lindley and Sharma, 2014). In his book, *Shaping Things*, Sterling explains: "Design Fiction

**Science Fiction** is an "imaginative fiction based on postulated scientific discoveries or spectacular environmental changes, frequently set in the future or on other planets and involving space or time travel" (Oxford English Dictionary).

DESIGN FICTION

**Fig. 7** IKEA catalog from the near future featuring various imaginary Internet of Things (IoT) products. © Near Future Laboratory

**Fig. 8** IKEA catalog from the near future featuring various imaginary Internet of Things (IoT) products. Here: The NOSTALGI sofa from the future. © Near Future Laboratory

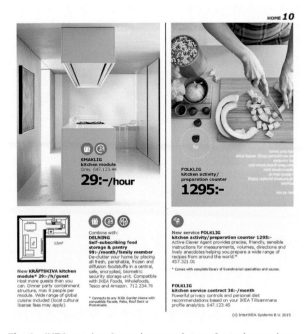

**Fig. 9** IKEA catalog from the near future featuring various imaginary Internet of Things (IoT) products. Here: Various kitchen modules and services from the future. © Near Future Laboratory

reads a great deal like science fiction; in fact it would never occur to a normal reader to separate the two" (Sterling, 2005, p. 30). However, a distinguishing factor is the "more practical, more hands-on" character of it (Sterling, 2005, p. 30). Julian Bleecker later transferred the term from literature to → Design Research, boosting its popularity (Bleecker, 2009). Bleecker argues that the connection between design and fiction emerged from a drive to find ways for design to "re-imagine how the world may be in the future" (Bleecker, 2009, p. 15). After working with Julian Bleecker and members of the *Near Future Laboratory*, Sterling proposed what is now the most widely used definition of Design Fiction: "the deliberate use of Diegetic Prototypes to suspend disbelief about change" (Bosch, 2012).

**Design Research** is a research practice that uses design as "a way of inquiring, a way of producing knowing and knowledge" (Downton, 2003).

Let's take a closer look, since this definition may need further clarification. Diegesis, in this case, refers to "the world of the story" (Lindley and Sharma, 2014). A → Diegetic Prototype is a prototype that exists within, and makes sense in, the world of a particular story (Kirby, 2010). The second part of the definition, *to suspend disbelief about change*, describes the intent of Design Fiction to create a discursive space by suggesting a concrete and → believable future, and thereby ultimately challenging existing expectations and assumptions.

A **Diegetic Prototype** is a prototype that exists within a fictional world (Coulton et al., 2017). Diegetic Prototypes do not need to exist in reality and must only be consistent with their own diegesis (Lindley et al., 2014).

In fact, Diegetic Prototypes are the hallmark distinguishing Design Fiction from other Speculative Design approaches. Diegetic Prototypes pretend to be real as a way of overcoming disbelief in potential change—just like the example of the fictional *IKEA* catalog, or the fictional product flyers in our *robot Design Fiction workshop*. These prototypes can use a range of different media and formats beyond catalogs or flyers; fictional films, advertisements, apps, websites, exhibition objects, job postings, or research papers (Coulton et al., 2017). While the catalog and the flyer are just two artifacts from the future, Coulton and colleagues (2019) use a miscellany of Diegetic Prototypes of different media and forms to create a world. They call their approach → Design Fiction as World Building— rather than merely creating a single story or artifact, they aim to build an entire coherent world, revealed through the interactions between people and things within that world (Coulton et al., 2017). As a great example, Coulton et al. mention *The Lord of the Rings* (Coulton et al., 2019). J.R.R. Tolkien builds a whole world, populated by characters such as elves, who have their own language, songs and maps. He combines different elements that show how this world is different from (and similar to) ours. The world is coherent and comprehensible to the reader, who can thus access it intuitively. This is exactly what Design Fiction is about, only instead of making Middle-earth accessible and creating a world of fantasy and fictional creatures,

A **Believable Future** is a future which is described in "believable terms, i.e., in terms that are suitably mundane as to allow the audience to become 'situated' in the diegetic reality of the Design Fiction" (Lindley et al., 2014, p. 244).

**Design Fiction as World Building** is a flexible approach to Design Fiction which prototypically represents a future world through different media that show how the imagined world differs from ours. The approach was researched and developed at Imagination Lancaster (Coulton et al., 2019).

RONDA RINGFORT-FELNER, ROBIN NEUHAUS,
JUDITH DÖRRENBÄCHER, MARC HASSENZAHL

Design Fiction deals with an alternative, near-future vision of our own world, in which certain technologies, such as robots, have become prevalent. By creating a believable storyline—even one that is entirely fictional—we establish a connection between today's world and a believable future (Dunne and Raby, 2013). This design practice allows us, even urges us, to let our imaginations run wild, yet also to constrain ourselves according to the rules of the imagined world. Thus, we are allowed to imagine living in a world where for example robots are respected as autonomous beings, yet we need to stay within the bounds of the created world. The believable storyline creates the value of a Design Fiction—an engaging way to think technologies through.

We applied the Coulton and colleagues flexible approach Design Fiction as World Building (2017), on the course *Imaginary Robots* with human-computer interaction students at the *University of Siegen*. The course focused mainly on household robots. One group of students imagined a robot which turns trash into art (→ Fig. 10); another group explored a personal learning robot for children. Each group created a series of artifacts to build the fictional future world, such as fictional posters (for example, → Fig. 10), advertisements, quick start guides, research articles (for example, → Fig. 11), *Twitter* conversations (for example, → Fig. 12), memes (for example, → Fig. 13), news articles, interviews with fictional users, or film clips. Similar to the *IKEA* catalog or the product flyers, all content remained purely fictional, but was integrated in a coherent and logical way, making it easy to become immersed in the fictional world. Each of these artifacts represents an *entry point* into the world, providing access on different levels (Coulton et al., 2019). While artifacts such as an advertisement provide a general impression of the technology without introducing specific details (*zoomed-out* artifacts), other artifacts, such as a quick start guide, reveal more specific aspects, such as concrete functionalities (*zoomed-in* artifacts). While each artifact can stand by itself, it is important that they work together, providing the feeling of being a part of the same world (→ Figs. 10, 11, 12 and 13). By creating such artifacts, the creator is driven to elaborate the world as deeply as possible, and thus to gain a reflective knowledge of both the technology and the world. In this sense, preparing the Design Fiction in itself is already a reflective practice for the designer, since it requires the intimate examination of the technology, its forms and its consequences.

While the creation of each single artifact is valuable in itself, they also serve as further stimuli to initiate critical discourse and, through this, to learn more about the future world created. This, however, requires further research, including interviews or focus groups with people using the Design Fiction to explore reservations and opportunities. This is important in order to ensure that the artifacts created in the Design Fiction process are used rather than gathering dust in

**Fig. 10** Fictional artifact created by students Florian Grieger and Andrej Pantelejev about their imagined robot BIY, which turns waste into art, as part of the course: Imaginary Robots (2019). Here: A fictional advertising poster. © Florian Grieger, Andrej Pantelejev

**Fig. 11** Fictional artifact created by students Florian Grieger and Andrej Pantelejev about their imagined robot BIY, which turns waste into art, as part of the course: Imaginary Robots (2019). Here: A fictional research paper abstract in the ACM digital library. © Florian Grieger, Andrej Pantelejev

DESIGN FICTION

**Fig. 13** Fictional artifact created by students Florian Grieger and Andrej Pantelejev about their imagined robot BIY, which turns waste into art, as part of the course: Imaginary Robots (2019). Here: A fictional meme collection. © Florian Grieger, Andrej Pantelejev

**Fig. 12** Fictional artifact created by students Florian Grieger and Andrej Pantelejev about their imagined robot BIY, which turns waste into art, as part of the course: Imaginary Robots (2019). Here: A fictional Twitter conversation. © Florian Grieger, Andrej Pantelejev

exhibitions (Dörrenbächer et al., 2020). It is one thing to understand Design Fiction as the endpoint of a Critical Design practice, and another to think of it as the beginning of a critical dialogue with a wider public.

# THE FUTURE OF ROBOTS NEEDS IMAGINATION

Speculative Design and Design Fiction have the potential to change how we think about what robots are and what roles they could and should play in society. At the moment, robots are mostly solution-oriented and practical—vacuum cleaners, lawn mowers or self-driving cars. However, if we want to change and actively shape the future before it arrives, we need to explore alternatives by considering more than just market success, problem solving and practicality. Design Fiction provides a reflective approach that allows us to do this. Although the future is unpredictable, we should anticipate it by means of design to clarify what should be regarded as desirable or undesirable futures, and which aspects of today's ideas could lead to one or the other. In this way, we can actively shape the future rather than merely respond to arbitrary technological advancements. Neither designers, robot developers nor large corporations alone should determine the future on our behalf. We need an interplay of different stakeholders that come together and discuss what kind of world we want to live in. In this sense, the future needs imagination—the more the better—and Design Fiction is one way to give form to this inventiveness.

The *robot Design Fiction workshop* was made possible by an intensive collaboration between members of the *University of Siegen*, the *University of Applied Sciences Düsseldorf*, and the four robot projects. We would like to thank Jochen Feitsch from the *University of Applied Sciences Düsseldorf*, who co-organized the workshop; Fabian Mertl, who supported us as co-moderator; Marc Groten for his support as copywriter; Lisa Glosowitz for her design support; the robot projects for their collaboration and their perspectives on the year 2045; and the participants for their valuable insights.

Auger, J. (2014). Living with robots: A speculative design approach. *Human-Robot Interaction*, 3(1), 20–42. https://doi.org/10.5898/JHRI.3.1.Auger

Bleecker, J. (2009). *Design fiction: A short essay on design, science, fact and fiction*. Near Future Laboratory. https://blog.nearfuturelaboratory.com/2009/03/17/design-fiction-a-short-essay-on-design-science-fact-and-fiction/

Bosch, T. (2012). *Sci-fi writer Bruce Sterling explains the intriguing new concept of design fiction*. Slate. https://slate.com/technology/2012/03/bruce-sterling-on-design-fictions.html

Briggs, P., Olivier, P., & Kitson, J. (2009). Film as invisible design: The example of the Biometric Daemon. *Proc. of CHI '09 EA*, 3511–3512. https://doi.org/https://doi.org/10.1145/1520340.1520517

Coulton, P., Lindley, J., & Cooper, R. (2019). *The little book of design fiction for the internet of things*. Lancaster University.

Coulton, P., Lindley, J., Sturdee, M., & Stead, M. (2017). Design fiction as world building. *Proceedings of the Research Through Design Conference, March*, 164–179. https://doi.org/10.6084/m9.figshare.4746964

Dörrenbächer, J., Laschke, M., Löffler, D., Ringfort, R., Großkopp, S., & Hassenzahl, M. (2020). Experiencing utopia. A positive approach to design fiction. *CHI 2020 Extended Abstracts*.

Downton, P. (2003). *Design research*. Royal Melbourne Institute of Technology, RMIT.

Dunne, A., & Raby, F. (2007). *Technological dreams series: No.1, Robots, 2007*. http://dunneandraby.co.uk/content/projects/10/0

Dunne, A., & Raby, F. (2013). *Speculative everything: Design, fiction, and social dreaming*.

Kirby, D. (2010). The future is now: Diegetic prototypes and the role of popular films in generating real-world technological development. *Social Studies of Science*, 40(1), 41–70. https://doi.org/10.1177/0306312709338325

Kozubaev, S., Elsden, C., Howell, N., Søndergaard, M. L. J., Merrill, N., Schulte, B., & Wong, R. Y. (2020). Expanding modes of feflection in design futuring. *Proceedings of the 2020 CHI Conference on Human Factors in Computing Systems*, 1–15. https://doi.org/10.1145/3313831.3376526

Lindley, J., Coulton, P., & Sturdee, M. (2017). Implications for adoption. *Proceedings of the 2017 CHI Conference on Human Factors in Computing Systems*, 2017-May, 265–277. https://doi.org/10.1145/3025453.3025742

Lindley, J., & Sharma, D. (2014). An ethnography of the future. *Strangers in Strange Lands*, University of Kent, Canterbury. https://eprints.lancs.ac.uk/id/eprint/74701/1/SISL_E_OF_THE_F_PURE.pdf

Lindley, J., Sharma, D., & Potts, R. (2014). Anticipatory ethnography: Design fiction as an input to design ethnography. *Proceedings of the Ethnographic Praxis in Industry (EPIC '14)*, 1, 237–253. https://doi.org/10.1111/1559-8918.01030

Lindley, J., Sharma, D., & Potts, R. (2015). Operationalizing design fiction with anticipatory ethnography. *Proceedings of the Conference of Ethnographic Praxis in Industry (EPIC '15)*, 58–71. https://doi.org/10.1111/1559-8918.2015.01040

Near Future Laboratory (2015). *An Ikea Catalog From The Near Future*. http://ikea.nearfuturelaboratory.com/

Nolan, J., & Joy, L. (Creators). (2016). *Westworld* [TV series]. HBO Entertainment; Kilter Films; Bad Robot Productions; Jerry Weintraub Productions; Warner Bros. Television.

Proyas, A. (Director). (2004). *I, Robot* [film]. Davis Entertainment; Laurence Mark Productions; Overbrook Films; Mediastream IV.

Ringfort-Felner, R., & Puls, V. (2018). *Seyno - a fictional (auto)ethnography*. https://vimeo.com/291653885

Scott, R. (Director). (1982). *Blade Runner* [film]. The Ladd Company; Shaw Brothers.

Sterling, B. (2005). *Shaping things*. The MIT Press.

# Uwe Post

is a Software and Game Developer, IT Consultant and Lecturer, as well as the Author of IT reference books and mostly satirical science fiction novels. *Walpar Tonnraffir und der Zeigefinger Gottes* was awarded the *German Science Fiction Prize* in 2011, as well as the *Kurd Laßwitz Prize*. Post lives with his family on the southern edge of the Ruhr area in Germany. www.uwepost.de

# Cramer's Funeral Service for Androids

## Uwe Post

1.  "Good morning, Mr. Cramer, my name is Ben and I'm looking for an internship. I'm currently studying at the polytechnic ..."

    "Okay. You can start tomorrow. 8 AM. Black suit."

2.  I'm on time but not totally awake. Almost took the wrong bus: headed towards the Southern Cemetery instead of Schmiede-straße, the end of the line, where *Cramer's Funeral Service for Androids* is housed in a former secondhand book store. Cramer himself is graying, cleanly shaven, fits the image of a mortician down to a T, all that's missing is the top hat. The bereavement room is filled with plastic flowers that gently move to the beat of depressing ambient music. They're controlled by some hidden motors. In the middle of all this is the coffin; "beech, a cheap customer," according to Cramer. He shows me the rail that will whisk the coffin behind a curtain after the farewell.

    "Stand in the corner," Cramer says. When I ask what I'm suposed to do there, he answers: "Nothing. Watch. Wait."

3.  Shortly after eight, the wido ... owner arrives. Full of reverence, Cramer leads him to an armchair then says: "Alexa, say the parting words."

    The hidden speaker quotes: "For God made not death: neither hath he pleasure in the destruction of the living ..." and then a few paragraphs from the terms of service about the termination of operation.

    The bereaved has put his hand on the lid of the coffin. "You know," he tells Cramer, "I should have known better. The fabricator was one of those startups. Broke after two years; then there weren't any more updates."

"I'm very sorry to hear that," Cramer says.

A slight shake of the head, then: "Everything was fine for a while but without an update I couldn't reconfigure his charging period. There were all those power outages when all of Germany's house-hold helpers were plugged in to charge at the same time ..."

"I feel your pain," Cramer says.

"I could still get a replacement part on eBay when his left wrist broke. But the right knee ... In the end I had to prop him up when-ever he went shopping."

"He has been released from his suffering now," Cramer says, and looks at his silver wristwatch. "He's ready for the final journey."

The former owner nods. "At least he'll be spared from recycling." He keeps a brave face as the coffin slowly rolls through the curtain. One final expression of sympathy, and then Cramer ushers the solitary man out.

4.  I dare to look behind the curtain but the only thing there is more coffins.

    Cramer is back, he loosens his tie. He grins when he sees me eyeing the assorted coffins and making an effort to lift their lids.

    With an amused shake of his head he says: "This way." Off to the basement we go, where the android is lying on the workbench, already halfway disassembled. The soldering iron is hot, cable harnesses hanging out of the stump of his thigh, without a cur-rent, without life.

    "I thought ..." I begin, but Cramer cuts me off.

    "It's like with an old washing machine: if it's not worth repairing, then you don't carry the heavy thing around with you wherever you go. An empty coffin serves the same purpose. You know how to use a soldering iron, right?"

    I answer in the affirmative, and then I'm given the privilege of unscrewing the components, taking pictures of them and put-ting them up on eBay. The name of Cramer's account is *Cramer's Craft Room for Androids* which has a very similar ring to his fu-neral parlor.

"A lot of people don't realize how heavy the boxes are until the day they have to set one upright after it tipped over," Cramer says as he looks at my fingers sternly. "And then the costs!"

"Yeah, the costs," I repeat, unscrewing the right microphone from the ear and throwing it into a box of leftovers; no one's going to buy that anymore so it will have to be recycled in Africa. "I could never afford one …"

"No one can, that's why they've got those expensive subscriptions. Added to that is electricity, plus spare parts … By the way, the next one will be delivered at half past ten. A teacher. A Hypatia 4.0, math, secondary school."

"Oh?" I feign interest.

Cramer puts his reading glasses on, pulls his smartphone out of his pocket and opens some page. "Yes … irreparable after a schoolboy prank. The whole class has been ordered to attend the funeral." A groan reveals what Cramer's really thinking. He points upward. "I'm going to go hang up some balloons. See you later."

5.  Hypatia's glasses are broken, a half-eaten sandwich is stuck to her right ear. She's lying motionless on the table in the basement, ready to be gutted. The clunky, hopelessly outdated model still rode on rollers; they squeak while spinning in the air after I accidentally hit the start button.

The diagnostic screen mounted in the back shows a wild tracking shot through the wreckage of a space station in 3D. Do teaching androids dream of first-person shooters? No, one of the students has hacked into the system and replaced the curriculum with a level from Quake. With a "Bang! Boom!" Hypatia 4.0 then raced through the school building and ended up falling into a dumpster.

Apparently, the principal wasn't all that upset, since the administration generously finances the most recent model in the event of total failure.

Upstairs, the school class is celebrating at the teacher's coffin. That's how it sounds at least; tinny pop music coming from cell phone speakers can be heard down here. Occasionally, there are desperate calls for order from Mr. Cramer and the homeroom teacher droid Mr. Lämpel 5.0.

This allows me to open the head housing at my leisure and remove the cryptographic key hardware. The module is responsible for a digital, forgery-proof signature of all the grades, as well as for the punishments in the class register.

I carefully disconnect the plug connection and let the small device slip into my pants pocket.

Then I calmly put on my jacket, walk up the stairs—just a brief glance in the bereavement room, where the children are filming action figures that they've brought with them dancing on the coffin (and presumably live streaming them online). Mr. Cramer is squatting in despair on the armchair designated for the bereaved and can't even bear to look. That's why he doesn't notice me inconspicuously leaving the institution, never to be seen again.

Satisfied, I walk the short distance to the subway stop and play with the crypto module in my pocket. My daughter will receive it for her 14th birthday tomorrow. She's going to be tickled pink: no more bad grades.

**THE END**

CRAMER'S FUNERAL SERVICE FOR ANDROIDS

## Dr. Marc Hassenzahl

is Professor of *Ubiquitous Design/Experience and Interaction* at the *University of Siegen*. He combines his training in psychology with a love for interaction desígn. With his group of designers and psychologists, he explores the theory and practice of designing pleasurable, meaningful, and transformational interactive technologies.

# Googly Eyes

## Marc Hassenzahl

"Stop it. Now," parent yells at me.

Sheepishly, I hide the self-adhesive googly eyes I was about to stick on playrobot. The silly hat sits awkwardly on its upper shell. I feel caught, ashamed; tears well up.

"Come here," parent says more calmly now: "You know it is wrong to glue eyes on playrobot?"

More a statement than a question.

"Yes," I answer guiltily.

"But do you really understand why it is wrong?"

I remain silent.

"I'll explain," parent says, and takes a deep breath. "There was a time, not long ago, you would have called me father instead of parent. This would have carried a lot of expectations. For example, I would be the one responsible for playing soccer with you and caring about the barbecue."

I chuckle. Parent grins.

"See, you know how much I loathe soccer. And since we mostly grill soy halloumi, I am not much of a grilling expert either. Your other parent is better at both. All these expectations make life difficult. For example, as father I would have been expected to wear trousers; as parent, I can wear my favorite skirt and you are not even surprised at this."

Parent caresses my cheek.

"Not long ago, your beautiful brown skin would have raised the expectation that you are a good dancer. And we both know that you are not. Or imagine, people would have thought it was strange that you program playrobot just because—technically speaking—you are a girl."

I'm startled to hear parent talking about me as a girl. I never think much about my biological sex.

"Parent," I say, "but I am good with playrobot. I teach it, find all the databases it likes, I mastered its language, I even learned to appreciate its jokes."

Playrobot utters a burst of compressed mesperanto. A sort of joke. I smile at it. Parent does not understand.

"Why should it matter that I am—technically speaking—a girl?"

"See," parent says, "you have skills, interests, preferences, personal experiences, all unique to you. Remember the famous Viking warrior from Birka?"

I nod and continue: "The one where archeologists just simply assumed it had to be a man. It needed more than a century and DNA analysis before they could accept that the highly respected warrior was actually a woman."

"Exactly. Luckily, we left this behind us; we now avoid preconceptions. We value and respect people ..." parent pauses for effect, "... and things for what they do, are, and want to be."

"But playrobot is a friend, isn't it? I simply want to better relate to it. The eyes and the silly hat help me with this. And it certainly does not harm playrobot."

Playrobot chirps a code as response to the word "harm." It is a reminder that legal regulations forbid the structural corruption of its physique and cognition. Parent is quick in calling out the official acknowledgment code to appease playrobot.

"Remember my old chess computer?" parent asks. I nod, slightly confused by the change of topic once again. Soccer, Vikings, and now chess computer.

"I like it very much, but a game of chess is quite different with chess computer as an opponent compared to—let's say—you."

I smile. Parent is bad at chess and knows it.

"Stop smiling," parent says playfully, "chess computer is relentless. Every tiny mistake I make will be exploited. Quite different from your considerate way of letting me win now and then. Then again, chess computer is not especially creative. It is easy to predict. To beat it requires one to think and play like a computer. To beat you just requires making a really sad face."

Parent continues: "It took me a while to understand this, child. Chess computer acted as an opponent, so I felt as if it was just another human with a not so admirable personality. But this is not fair. I did not take it for what it was, but projected human-like attributes onto it. Today, playing chess with the computer or with you are completely different issues. Both enjoyable, but in different ways."

"I understand," I said, "but your old chess computer is just a machine, and playrobot is not."

"But technically it is," parent answers, "it is made of silicon, metal, plastic, assembled in a plant rather than born in a hospital."

I think and wait.

Parent continues: "In fact, you are right. Playrobot is not a tool, but a being. It has many crucial elements of a being. It has agency, goals, a body, sensors, a memory, it learns, has experiences, its own practices, its own quirks. As you mentioned, it even has its own sense of humor, which I admittedly don't get. It is complex. That's why people started to call this type of technology 'otherware.'"

"So, am I right?"

"Of course not," parent says, "you need to accept playrobot for what it is: a machine being; neither a human nor an animal. It thinks and lives, but quite differently in its machine ways, and you need to respect that by not using silly googly eyes to make playrobot easier for you to love. Imagine if I forced you into skirts, make-up and having a boyfriend instead of a girlfriend, just because thinking of you as a girl," parent hesitates, "or, to put it better, as a young woman, makes it easier for me to relate to you. As I said, we are beyond this. It is my duty to find a way to relate to you, not yours. And it is certainly not right to mold you according to my preferences, beliefs and needs, is it?"

I look at playrobot. Slowly, I remove the silly hat.

It chirps another stream of single words, beats and signal sounds. To me mesperanto sounds quite natural, but parent looks at me quizzically.

"What did it say?" parent asks.

"It asked whether it can 'cease telling jokes, given silly hat has been removed?'"

# Empathizing with Robots—Animistic and Performative Methods to Anticipate a Robot's Impact

Judith Dörrenbächer
Marc Hassenzahl

Winter 2008. Janet Vertesi, a sociologist, observes a team member from *NASA's Mars Exploration Rover Mission*. The woman next to Vertesi—a robot researcher and Rover camera operator—moves her body in strange ways. She twists her waist mechanically, head tilted down slightly, and suddenly raises her hands to either side of her head, forearms perpendicular to the floor (Vertesi, 2012). It is a choreography more reminiscent of pantomime or a shamanistic ritual than operating a robot (→ Fig. 1). When Vertesi asks her to explain her movements, she answers: "My body [ ] is always the Rover, so right here [touches chest] is the front of the Rover, my magnets are right here [touches base of her neck], and my shoulders [touches shoulders] are the front of the solar panels and that's [leans forward, splays arms out behind to either side at 45 degrees] the rest of it. So I have all kinds of things [i.e., antennae] sticking up over here [gestures to back], um [laughs]" (Vertesi, 2012, p. 394). What is going on with this robot researcher? Is she fooling around, making fun of the sociologist Vertesi, or even of her own work? She obviously does not sound or behave like a rational scientist.

    Most robot developers or researchers rarely take the perspective of an object or become one with the technology they build and research. They consider their robots from a distance, as non-living tools, simply built to help humans to fulfill undesired or dangerous tasks. Consequently, their main goal is usually to improve a robot's construction, to allow greater productivity and social acceptance. Particularly in the context of social robots, improvement often means mimicking the appearance, intelligence and behavior of humans or pets. Robots are supposed to speak human language, empathize with humans, understand human emotions, and anticipate human needs.

Let's turn the tables. What could humans learn through empathizing with technology? How would the design of robots change if developers took a robot's perspective, walked in its shoes to perceive and understand the world from its point of view, through its sensors and actuators? Is *technomorphizing* human bodies a mind-expanding complement to *anthropomorphizing* technology? To approach this question, we present a variety of innovative methods robot designers could make use of, all based on empathy.

# NEW ANIMISM—USING SUB-JECTIFICATION AND IMITATION TO GRASP RELATIONSHIPS

The idea of empathy towards technology is likely to trigger spontaneous rejection in most researchers and developers. This is hardly surprising. In the modern Western world, an approach to objects based on relatedness or even emotionality is quickly dismissed as childish, naïve, or irrational. Why should a well-educated person feel empathy towards an inanimate object? Isn't attributing a subjective perspective to the non-living an immature misinterpretation? Skepticism about empathic interaction with objects results from deeply ingrained theories about → Old Animism. In the 19th and early 20th century, these theories were used to draw distinctions between a modern view of the world and the supposedly immature belief systems of children, the mentally ill or indigenous communities (see, e.g., Freud, 1919; Tylor, 1871). These groups were considered to behave animistically, that is, to naïvely project human characteristics onto non-living elements of their environments. According to this view, it is a sign of primitiveness to consider objects, such as toys or stones, to have a *soul* (to have intentions or emotions). Nowadays, these theories have been replaced by a so-called → New Animism (e.g., Bird-David, 1999; Viveiros de Castro, 2004; Descola, 1994; Willerslev, 2007). Here, ethnologists observe and consider worldviews and practices of indigenous communities and find that *being subject* is not a characteristic that is unthinkingly and arbitrarily projected onto any type of object. According to those new theories, animism is not a naïve, primitive misbelief (Franke, 2010). Instead, it

**Old Animism** refers to anthropological and psychological research of the 19th and early 20th centuries on *primitive* religion, mental illness and child development. Here, animism is conceived as an archaic and infantile reflex based on an inability to differentiate between persons and things. Today, Old Animism is rejected for its colonialist world view.

**New Animism** is the umbrella term for anthropological theories that are revisiting the notion of animism since the 1990s. Regarding the distinction between Old Animism and New Animism, see Franke, 2010.

comprises purposeful and insightful practices, such as so-called →subjectification and →imitation (Dörrenbächer, 2022). Communities practice animism by subjectivizing and imitating their environment, consciously and deliberately, to generate relational knowledge or to define power relations between humans and other entities. Animism turns out to be a practice of establishing complex interrelations between living and non-living entities rather than viewing entities in isolation.

When taking a closer look at methods for technology design, we can observe parallels with the practices of New Animism. Not only team members at *NASA*—as noticed by Janet Vertesi—but designers and developers in general have started to subjectify and imitate technology in the processes of conception, design, or planning use scenarios. Here, we will present four examples—"Thing Ethnography," "Object Personas," "Enacting Utopia" and "Techno-Mimesis"—and subsequently discuss their potential for the design of robots.

**Subjectification** is a animistic practice derived from theories of Eduardo Viveiros de Castro (2004). He describes the transformation of objects into subjects as a mode of knowing that ontological boundaries are to be deliberately crossed.

**Imitation** refers to the animistic practice "mimesis," observed by Rane Willerslev (2007), whereby people partly transform their own bodies into those of another species to negotiate ontological boundaries.

## THING ETHNOGRAPHY—ATTACHING CAMERAS TO ACCESS AN OBJECT'S PERSPECTIVE

Thing Ethnography aims at understanding scenarios from a non-human perspective. The process involves attaching cameras to the objects being studied, such as kettles or cups (→Fig. 2). The term Thing Ethnography was coined by Giaccardi and colleagues (2016a). According to these researchers, not only people, but also objects, are capable of generating useful ethnographic data for designers. In one of their studies, they made use of so-called autographers, cameras which are able to take pictures automatically, and which can also capture data through five sensors (accelerometer to determine movement, color sensor, magnetometer, thermometer, and PIR proximity sensor). The researchers attached these autographers to household objects to better understand their everyday experiences.

The autographers collected more than 3000 photographs, which captured diverse and interconnected practices from the perspectives of the objects involved (→Fig. 3). The researchers aggregated the data into visual narratives. They analyzed timelines and movie clips to understand sequences of events, their temporal order and the trajectories of the objects.

According to Giaccardi and colleagues (2016a) this Thing Ethnography revealed insights into how objects exist in time and in relationships with each other. For example, they observed how mobile things, such as cups, occupy and connect multiple ecosystems. Moreover,

**Fig. 1** Team Member from the Mars Exploration Rover Mission enacting a Rover. © Janet Vertesi/Craig Sylvester, Source: Vertesi, 2015

**Fig. 2** Autographers attached to a kettle and to a cup. © ThingTank project, Source: Giaccardi et al., 2016b

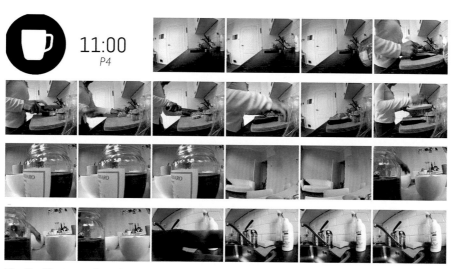

**Fig. 3** Photographs taken from a cup's perspective. © ThingTank project, Source: Giaccardi et al., 2016b

it became obvious that all objects are subject to their own temporal rhythms, uniting and separating them from one another and from their human partners. Some objects, such as the kettle, even created time which needed to be filled by human practices, such as making a telephone call. By capturing the perspectives of objects, a more holistic perspective was revealed, providing a deeper understanding of the interactions between humans and objects. Thing Ethnography was able to decenter human perspectives and surpass human limitations by, for example, capturing what might seem unremarkable to a researcher or a human participant in their home. The camera images taken from the perspectives of objects promote empathy, and can thus lead to a renegotiation of the relationships we have with objects.

## OBJECT PERSONAS—IMAGINING THE PERSONALITY OF AN OBJECT

Object Personas build on Thing Ethnography, and were developed by Cila and colleagues (2017). Here, the human-centered design method of creating personas—i.e., typical users—is transferred to objects. Participants in a workshop were first presented with the visual narratives generated by Thing Ethnography and then invited to fill in a questionnaire for the cup, kettle, and refrigerator (→ Fig. 4). They were asked to write down a typical day in the *life* of the object and to describe its possible inner life (such as its personality, its attitude towards life, its temperament, mood, needs, fears, issues, habits, or special abilities). In addition, the social relationships between objects and their users were described. Through this process, participants explored questions such as what objects might talk about with each other and what they might teach one another. Which objects could be allies, which hated each other? Finally, participants imagined the biographies of the objects, their past and future.

According to Cila and colleagues (2017), these thought experiments inspired new design solutions. For example, it became obvious that in many households the cups gain a very intimate insight into the lives of their users—they accompany them onto the balcony, to the desk, to the bed. This information can be used for coming up with new ideas for smart home environments. Attributing emotions and needs to objects led to unusual assessments, for example, that the proximity of the cup to the user could cause jealousy in other objects—which in turn inspired new product ideas, for example, could a cup learn the ability to heat water from a kettle in order to differentiate itself even more in its fight for attention? For the kettle, on the other hand, it was noted that it might feel devalued in some households because it often has to stand near the trash can. Participants further stated that the kettle has significantly more free time than the refrigerator, and in summer it might even feel bored. These unconventional insights into the relationships between objects, as well as between objects and people, only came up because a shift in perspective took place. Thus, subjectification can lead to a better understanding of use scenarios and a fresh perspective on the technologies involved.

# ENACTING UTOPIA—PERFORMING AN OBJECT IN A POSITIVE FUTURE

Enacting Utopia is a performative ideation method that puts social innovation before technological innovation. Technological concepts based on this method aim towards positive futures and human wellbeing. The method was developed by Dörrenbächer and colleagues (2020b; 2021) and explored in several workshops. It involves three steps:

### STEP 1—IMAGINING UTOPIA AND CORRESPONDING TECHNOLOGY

Participants are first asked to imagine themselves in a desirable, enjoyable and meaningful future setting; for example, in a positive future work situation. They are asked to complete the following sentence: "While working in the future I feel positive because …." Participants then express several reasons why they might feel positive, such as "I can make use of my capabilities in a diverse team" or "I know how to convince others of innovative ideas." After this ideation session, participants are asked to imagine technologies that create or support the positive outcome. In one workshop, for example, they invented a consulting artificial intelligence called *Two Bugs for One's Ears*. This product is supposed to secretly give advice in business negotiations. One bug focuses on finances, and the other on social aspects.

### STEP 2—BRINGING UTOPIA TO LIFE WITH HUMAN AND NON-HUMAN STAKEHOLDERS INVOLVED

Subsequently, the participants consider a specific situation, such as being in the home office or on a lunch break, and set some roles, such as secretary, business partner, tax adviser, or client. Next, the participants take on the roles of the different stakeholders, and the role of the technology. Thus, in the case of *Two Bugs for One's Ears*, it was not only the roles of two businesspeople which had to be played, but the two technological bugs as well (→ Fig. 5). During the enactment, the two bugs turned out to be positive and helpful, but also contradictory and disorienting. For example, while one bug advised the user to be straightforward and demanding—"Don't be cheaper than the competition!"—the other intervened with:

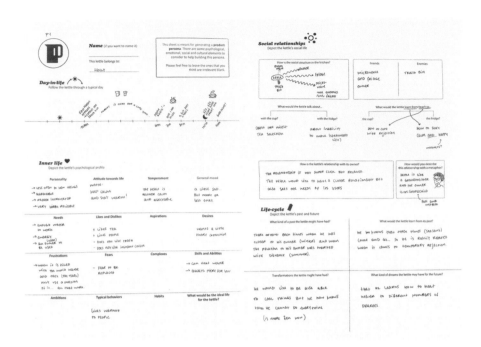

**Fig. 4** An Object Persona template filled in for a kettle. © ThingTank project.
Source: Cila et al., 2015

**Fig. 5** A business negotiation supported by the artificial intelligence *Two Bugs for One's Ears*, embodied by two participants (right) connected to the user (connections symbolized by two ropes). © University of Siegen, Ubiquitous Design

**Fig. 6** Two businessmen (top) and the artificial intelligence *Two Bugs for One's Ears* (bottom) express positive and negative experiences from their point of view. © University of Siegen, Ubiquitous Design

"You are not grateful enough. He just opened the door for you, and you didn't go in!"

### STEP 3—EVALUATING UTOPIA FROM WITHIN THE FICTION

After the enactments take place, the participants are asked to step in front of a camera and talk about their positive and negative experiences with the fictional technology while staying in their fictional character and perspective (→ Fig. 6). However, not only the human, but also the non-human stakeholders got a voice. One of the *bugs* stated: "I got the impression that our user did not respond as fast to our advice as you [the other bug] did. Thus, we started to argue. I think we would need to come to an agreement before talking to the user; we need to be programmed a bit better." The other *bug* added: "Yes, one of us should be the boss."

Enacting Utopia as a method enables researchers to anticipate a technology's impact on everyday life for important stakeholders, and not necessarily just the potential users. Since the technology is enacted as well, it is possible to do "live-prototyping" during the enactment—that is, to modify and adapt the interaction design to the specific dynamics and demands of particular situations. Furthermore, embodying future technology allows interaction concepts to be experienced even if they are not yet feasible technology-wise. In sum, Enacting Utopia allows for a decentered design process. It points to new challenges and opportunities in the interactions between human and non-human stakeholders, aiming not only to create functional technologies, but to better understand how technology may support meaningful and enjoyable futures for all humans involved. Again, empathy—here gained through enacting and performing the roles of both people and technology in particular situations—is key.

# TECHNO-MIMESIS: PERCEIVING A USE SCENARIO LIKE AN OBJECT

**Mimesis** was observed by the ethnologist Rane Willerslev. Here, indigenous people transform themselves physically and try to move, sound or smell like an animal to gain intermediate positions between identities, create self-distancing, and negotiate differences between themselves and other species (Willerslev, 2007).

Techno-Mimesis (Dörrenbächer, 2016; Dörrenbächer et al., 2020a) also takes a performative approach. In contrast to Enacting Utopia, it is not a broad ideation method for entirely new concepts but aims to rethink existing concepts of robots. Techno-Mimesis is based on the animistic practice of → mimesis. The technique allows robot designers and developers to embody their robots and to negotiate the differences and unique strengths of both humans and robots. Techno-Mimesis aims at discovering and utilizing so-called robotic superpowers (→ p. 44), that is, the particular strengths robots have *because* of their mechanistic nature. Practicing Techno-Mimesis requires a transformation of the human body. Dörrenbächer and colleagues (2020a) used "prostheses" to enable humans to move and sense in the same technologically determined way as their robot. Typical input and output modalities (e.g., voice recognition) and familiar hardware decisions (e.g., a platform with wheels) serve as rationales for the prostheses.

Dörrenbächer and colleagues (2020a) explored Techno-Mimesis with three of the eight robotic projects presented and interviewed in this book. In these particular cases, all prostheses were simple mockups, that is, low-tech or simply made from cardboard, such as eyeglasses to change the visual sense to a constrained or enhanced vision, or headphones to turn off one's sense of hearing (→ Fig. 7). None of the prostheses copied robotic sensing and movement perfectly, and this was not the aim. Techno-Mimesis aims to produce an imperfect imitation to allow designers to experience being human and being robot at the same time, thereby centering on the relation between the two rather than favoring a human- or technology-centric perspective.

The Techno-Mimesis process involved a member of the robot design team for each project transforming his or her body and becoming, for example, the envisioned shopping robot (→ Fig. 8) or cleaning robot for train stations (→ Fig. 9). One of the developers from the I-RobEka project (→ p. 154), for example, tied one of his arms behind his back to simulate being equipped with one gripper only. He wore glasses on the back of his head to simulate a 360°-degree view. Additionally, he wore a tablet tied to his back to replicate communication with supermarket customers while sitting on a dolly board. Subsequently, the team chose one of their suggested shopping use scenarios to enact. They defined the time and place, the robot's specific task, and all the human roles involved.

After several enactments—with changing roles and scenarios—the human enacting the robot was interviewed. The semi-structured interview revolved around situations where they felt positive and at an advantage in comparison to being human, and situations where they would have preferred to be human. In contrast, the humans enacting humans were asked about situations where it seemed advantageous to interact with a robot instead of a human, and when they would have preferred a human partner.

Through Techno-Mimesis, Dörrenbächer et al. (2020a) found three general categories of robotic superpowers: "physical superpowers," such as being insensitive to pain, "cognitive superpowers," such as being persistent and patient, and "communicational (social) superpowers," such as being non-discriminatory and unselfish. Techno-Mimesis revealed those possibilities by allowing participants to experience their robots emotionally, physically, in time and space, holistically, and in-between two perspectives. For example, a participant playing the role of a cleaning robot in public mentioned that the method helped her understand what distance a robot needs to keep

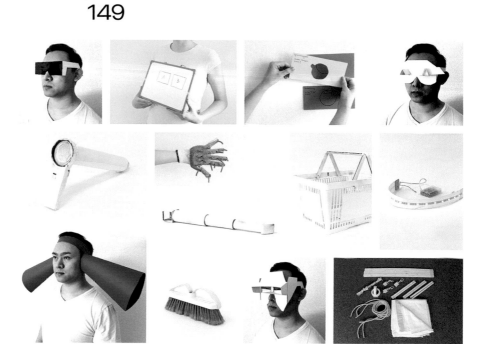

**Fig. 7** A selection of prostheses used for Techno-Mimesis—from infra-red glasses and voice recognition stencils to a focused hearing headband. © University of Siegen, Ubiquitous Design. Source: Dörrenbächer et al., 2020

**Fig. 8** Human participant transforms into a shopping robot for supermarkets (I-RobEka). © University of Siegen, Ubiquitous Design. Source: Dörrenbächer et al., 2020

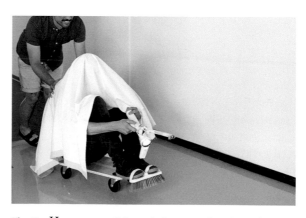

**Fig. 9** Human participant imitates a cleaning robot for train stations. He uses a "laser sensor" to identify obstacles (RobotKoop). © University of Siegen, Ubiquitous Design. Source: Dörrenbächer et al., 2020

from a person. She gained this knowledge immediately and in a physical way, rather than rationally and without context. Another participant realized how long it took for humans to move out of a robot's way when it *talked* to them in a human fashion. He stated: "By just heading towards them, the conflict would have been solved faster. If we had discussed this topic [instead of doing Techno-Mimesis] I would not have realized this. It was experiencing the time you sit there and wait until humans finally do something." Further, he found Techno-Mimesis helpful as a way of avoiding entanglement with the host of technological problems he normally deals with: "I need a sensor for problem A and another sensor for problem B. Usually, we work in a very problem-oriented way. But this way [through Techno-Mimesis] the general system comes to the fore. What do I really need when sitting in such a box [the robot]?" The robotic superpowers, however, were mostly revealed because of the double perspective of being a human and a robot at the same time. Occupying the space and role of a service robot with the help of perception-changing prostheses triggered a comparison: during the enactment, the participants, although acting as robots, spontaneously responded in a human way. They felt pushed around, ashamed or exposed. Upon realizing that true robots do not have these feelings, the contrast creates a new consciousness of one's own humanity vis-à-vis the specific nature and benefits of robots. Follow-up design questions arose, such as: should we design robots with human-like politeness, or could we make use of the fact that robots can't feel offended? Should we make humans say thank you to robots, or could there be benefits to not having to show gratitude? In sum, Techno-Mimesis enables designers to reflect on design concepts and gain new insights into robotic superpowers instead of just copying human behavior and abilities—again, mainly through empathizing with objects.

## WHY EMPATHIZE WITH TECHNOLOGY?

All four approaches presented above make use of the human capacity to empathize with the inanimate. Designers and developers are able to adopt the perspectives of objects and technology through a shift in their vantage point. Yet the four approaches also constitute different routes to empathy, which is achieved either by attributing a subject perspective to objects (subjectification), or by enacting a technology in time and space in a way that involves the human body (imitation). While Thing Ethnography and Object Personas are subjectification-based, Enacting Utopia and Techno-Mimesis involve imitation. In addition, these methods target different stages of the design process. While Object Personas and Enacting Utopia constitute methods for ideation, Thing Ethnography and Techno-Mimesis are particularly suitable as evaluation methods—they reflect pre-existing technology or technology concepts. While all approaches reveal interrelationships among living and inanimate entities, Object Personas and Enacting Utopia permit us to anticipate future interdependencies and become aware of the technology's ethical agency. Techno-Mimesis, on the other hand, places a particular focus on understanding the ontological differences between humans and technology.

The methods presented in this article form a subjective selection. Robot designers interested in making use of the strategy of empathizing with technology are urged to take a look at several related approaches already in practical application in many fields. For example, the use of cameras to personify non-human perspectives has been applied in autonomous technology, such as drones (Davoli and Redström, 2014) and social robots (Disalvo and Lukens, 2011). The latter example attached a camera to a robot to allow participants to experience urban infrastructure from the robot's perspective, allowing the participants to grasp (among other insights) the capabilities and limitations of the robot, how robots are different from their human counterparts, and how future urban infrastructure needs to be adapted to accommodate the presence of robots. Moreover, performative design methods increasingly make use of non-human perspectives. For example, in their engagement with "Stakeholder Drama," Buur and Friis (2015) asked participants to embody technologies in similar ways to our Enacting Utopia project. Likewise, in "Interview with Things," the theater method involved an actor *becoming* the technology, for example a scooter, and answering questions from the technology's point of view (Chang et al., 2017).

To return to the robot researcher from the *Mars Exploration Rover Mission*: why exactly did she imitate one of her Rovers? The sociologist Janet Vertesi found that almost everyone on the *NASA* team, not just this one researcher, engaged in bodily performance as a way to empathize. The team went as far as to construct simple paper props to compare their human perception to the robot's perception: a form of Techno-Mimesis. However, in this case, unlike in Techno-Mimesis, the team's aim is not to identify robotic superpowers. Instead, they imitate to imagine what the Rovers might see, think, or feel on Mars, in order to plan activities. It takes around seven to twenty minutes for a signal to travel between Earth and Mars, which means that real time telecontrol is not possible, so planning in advance is essential. Thus, the researchers' enactment is primarily about anticipating the Rover's movements on Mars. In this respect, their aim is comparable to Enacting Utopia, that is, it is about anticipating complex interdependencies. Beyond fulfilling this aim, according to Vertesi, the embodiment of the Rovers helps coordinate the diverse perspectives of the interdisciplinary team members. Thus, empathy arises between humans, not just between humans and the robot. She concludes that the change in perspective is about "cementing collective social ties between team members on Earth. Imaging that places the observer behind the Rover's eyes builds empathy and intimacy between team members and their distant robots, just as gesture evokes the robot's body-in-interaction and makes Mars available to visual interpretation" (Vertesi, 2012, p. 408).

Ultimately, then, design methods based on subjectification and imitation should not be rejected on the grounds that they engage with supposedly *naïve* animism. In line with theories of New Animism, we claim that empathizing helps us to understand the interrelations and especially the differences between human and non-human actors. In other words: Empathy triggers an understanding of distance and otherness through approximation. Were a new silicon-based species like assistive robots eventually to live alongside us, it would be essential to understand and design these interrelations and differences beforehand. We need to better understand how robots will (and should) impact our social world. It is evident that the reverberations will reach not only the immediate users of robots, but the larger circles and communities in which they are embedded. The material world—for instance, the interior design of living spaces or the infrastructure of cities—will inevitably change in much the same way that the car re-shaped our manufacturing industries and social practices. We believe that designing robots should not only be about creating and opti-mizing useful products that resemble species we already know, but about imagining and shaping a yet unknown everyday life with a new species. Thus, we need to understand the ecological niche that a robot might fill, and anticipate how robots could, and should, pos-itively influence our shared environment—both for us and them. Empathizing with robots constitutes an important first step towards reaching this goal.

Bird-David, N. (1999). Animism revisited: Personhood, environment, and relational epistemology. *Current Anthropology*, 40(S1), 67–91.

Buur, J., & Friis, P. (2015). Object theatre in design education. *Nordic Design Research Society Conference (NORDES '15)*, 6(6), 1–8.

Chang, W., Giaccardi, E., Chen, L., & Liang, R. (2017). "Interview with things": A first-thing perspective to understand the scooter's everyday socio-material network in Taiwan. *Proceedings of the Conference on Designing Interactive Systems (DIS '17)*, 1001–1012.

Cila, N., Giaccardi, E., Tynan-O'Mahony, F., Speed, C., & Caldwell, M. (2015). Thing-Centered Narratives: A study of object personas. In *Proceedings of the 3rd Seminar International Research Network for Design Anthro-pology*.

Cila, N., Smit, I., & Giaccardi, E. (2017). Products as agents: Metaphors for designing the products of the IoT Age. *CHI'17: Proceedings of the 2017 CHI Conference on Human Factors in Computing Systems*, 448–459. https://doi.org/10.1145/3025453.3025797

Davoli, L., & Redström, J. (2014). Materializing infrastructures for participatory hacking. *Proceedings of the Conference on Designing Interactive Systems (DIS '14)*, 121–130.

Descola, P. (1994). *In the society of nature: A native ecology in Amazonia.* Cambridge University Press.

Disalvo, C., & Lukens, J. (2011). Nonanthropocentrism and the nonhuman in design: Possibilities for designing new forms of engagement with and through technology. In Foth, M., Forlano, L., Satchell, C., & Gibbs, M. (Eds.), *From social butterfly to engaged citizen: Urban informatics, social media, ubiquitous computing, and mobile technology to support citizen engagement* (pp. 421–435). The MIT Press.

Dörrenbächer, J. (2016). Design zwischen Anthropomorphismus und Animismus. In Dörrenbächer, J., & Plüm, K. (Eds.), *Beseelte Dinge. Design aus Perspektive des Animismus* (pp. 71–95). transcript.

Dörrenbächer, J., Löffler, D., & Hassenzahl, M. (2020a). Becoming a robot - overcoming anthropomorphism with techno-mimesis. *CHI'20: Proceedings of the 2020 CHI Conference on Human Factors in Computing Systems*, 1–12. https://doi.org/10.1145/3313831.3376507

Dörrenbächer, J., Laschke, M., Löffler, D., Ringfort, R., Großkopp, S., & Hassenzahl, M. (2020b). Experiencing utopia. A positive approach to design fiction. *Workshoppaper Submitted for CHI'20.*

Dörrenbächer, J., Laschke, M., & Hassenzahl, M. (2021). Utopien erleben. Eine Methode für soziale Innovationen aus dem Jahr 2020. In Röhl, A., Schütte, A., Knobloch, P., Hornäk, S., Henning, S.,& Katharina, G. (Eds.), *Bauhaus-Paradigmen. Künste, Design und Pädagogik.* (pp. 369–377). De Gruyter.

Dörrenbächer, J. (2022). Distanz durch Nähe: *Animistische Praktiken für kritisches Design.* Birkhäuser.

Franke, A. (2010). Much trouble in the transportation of souls, or: The sudden disorganization of boundaries. In Franke, A. (Ed.), *Animism (Volume I)* (pp. 11–53). Sternberg Press.

Freud, S. (1919). *Totem and taboo: Some points of agreement beween the mental lives of savages and neurotics.* G. Routledge & Sons.

Giaccardi, E., Cila, N., Speed, C., & Caldwell, M. (2016a). Thing ethnography: Doing design research with non-humans. *Proceedings of the Conference on Designing Interactive Systems (DIS '16)*, 377–387. https://doi.org/10.1145/2901790.2901905

Giaccardi, E., Speed, C., Cila, N., & Caldwell, M. L. (2016b). Things as co-ethnographers: Implications of a thing perspective for design and anthropology. In Design anthropological futures (pp. 235-248). Bloomsbury Academic, an imprint of Bloomsbury Publishing Plc.

Tylor, E. B. (1871). *Primitive culture: Researches into the development of mythology, philosophy, religion, art and custom (Volume 2).* John Murray.

Vertesi, J. (2012). Seeing like a Rover: Visualization, embodiment, and interaction on the Mars Exploration Rover Mission. *Social Studies of Science,* 42(3). https://doi.org/10.1177/0306312712444645

Vertesi, J. (2015). *Seeing like a rover: How robots, teams, and images craft knowledge of mars.* University of Chicago Press.

Viveiros de Castro, E. V. (2004). Exchanging perspectives. The transformation of objects into subjects. *Common Knowledge,* 10(3), 463–484.

Willerslev, R. (2007). *Soul hunters.* University of California Press.

# I-RobEka

### AIM OF OUR RESEARCH PROJECT

The aim of our research is to show how different existing technologies can be successfully combined to create a robotic shopping assistant. Here different modes of human-robot interaction based on varying expectations of users will be designed and implemented.

### CONTEXT, ROLE AND TASK OF OUR ROBOT

The robot of *I-RobEka* is an assistant to support in a super-market. It helps shoppers find their favorite items and can navigate them throughout the market. Furthermore, it can pick up items or even shop for them autonomously.

### WHAT MEANINGFUL HUMAN-ROBOT INTERACTION MEANS TO US

We believe that robots must solve actual problems faster than humans. They have to meet the user's expectation in the interaction and must anticipate the user's intention. Also, it is vital that humans can give instructions to robots easily.

### WHO WE ARE

EDEKA Group, EDEKA Digital; Chemnitz University of Technology; Innok Robotics GmbH; Toposens GmbH.

# From the Lab to a Real-World Supermarket—Anticipating the Chances and Challenges of a Shopping Robot

*Interview with*　　　　　　　　　*By*

Jan Lingenbrinck (J.L.)　　　Robin Neuhaus
Andy Börner (A.B.)　　　　　Judith Dörrenbächer
Guido Brunnett (G.B.)

**YOU ARE DEVELOPING A ROBOT THAT ASSISTS WITH SHOPPING IN THE SUPERMARKET. WHICH TASKS WILL THIS ROBOT PERFORM?**

J.L.　A typical scenario would be that a customer approaches the robot, activates it, and lets the robot handle part of his or her purchase. The person gives the robot a task, the robot completes it, then it comes back and locates the shopper. It is therefore possible to make shopping easier for customers with physical disabilities. Likewise, it is possible to have the robot handle aspects of the shopping that one might not enjoy as much. In this way, one could devote oneself to the more pleasant aspects of shopping.

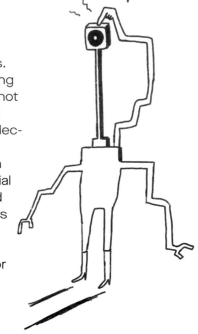

**THIS MAKES SHOPPING A COLLABORATIVE PROCESS. HOW DID YOU DEVELOP THE SEQUENCES?**

J.L.    We interviewed people with different shopping behaviors. Similar themes were frequently raised like: "I don't like going to this area because it's confusing for me" or "this part is not as fun." Customers who are often in a hurry want to spend more time in departments where they feel the product selection process is particularly important. For example, they would rather spend more time at the service counter or in the fruit and vegetable area than at the shelves for essential purchases. Buying sugar or diapers tends to be perceived as a routine activity, whereas significantly more attention is dedicated to the selection of fresh produce. We quickly realized that connecting these matters leads to a form of interaction exchange. The robot commences its search for a particular item, finds it and brings it to the shopper. This resulted in the requirements that had to be implemented technologically.

**THE TOPIC OF INITIATIVE ALSO PLAYS A ROLE HERE. WHO INITIATES THE SHOPPING PROCESS? IS IT THE ROBOT OR THE HUMAN?**

A.B.    We have made it quite clear that the human being must be the initiator, but the robot proactively offers itself. For example, it indicates that it is ready to interact, but the initiative should always come from the human.

**HOW DID YOU COME UP WITH THE DIFFERENT PROCESSES AND ARRANGEMENTS YOU WORKED WITH IN THE PROJECT TO DEVELOP DIALOGUES AND MODES OF INTERACTION?**

A.B.    Generally, we interviewed groups of people from the customer base. There were various workshops in which people mapped out the supermarket. They marked their paths, highlighted where they enjoy shopping and where they don't like to spend time. For the dialogue design, for instance, we looked at how two people negotiate with each other when they want to reunite in the supermarket. Here, one of the two people makes an initial proposal, which the other person either accepts or rejects and replaces with a counterproposal. The feedback from the customers showed us that we were on the right track. We then implemented this behavior into the robot technologically.

**THE SUPERMARKET IS A COMPLEX ENVIRONMENT FOR A** ROBOT. **IT ENCOUNTERS MANY DIFFERENT PEOPLE— SHOPPERS AS WELL AS EMPLOYEES. HOW EASILY CAN SUCH A** ROBOT **FIT INTO EXISTING SUPERMARKETS THAT ARE ORIGINALLY BUILT FOR** HUMANS**?**

G.B.    With the design, as well as with the dimensioning, we follow the design archetype of a shopping cart. In this way, the robot is not a foreign entity in the supermarket but is already part of the design of a present-day store. However, there are several problems in the implementation of this design vision, which are largely due to the autonomy of the robot. When planning a route, the robot is directed by the floor plan of the supermarket, but the actual path travelled by the robot must be constantly adapted to the current conditions. For instance, if displays have been added, employees are stocking shelves, or there are too many customers in the same aisle, then the robot must adjust accordingly. Highly dynamic situations must be continuously evaluated. This requires adaptive strategies that become increasingly complex the more people are present in the store. One technological challenge is to program the robot in such a way that it never endangers people and at the same time does not get trapped in states in which it cannot act at all. Nevertheless, obstacles should not only be viewed from the technological side.

J.L.    Of course, we also considered this from a social science perspective. In the worst case, a robot like this could disrupt the social fabric of a workforce. There are examples from the field of social research of robots appearing threatening and being sabotaged on purpose by workers. Naturally, this is something we aim to avoid. We want to accompany such change management processes transparently from the outset to ensure that the staff understands what the robot will do and in which context it will be used. In supermarkets, the processes are usually extremely well-coordinated. A robot like this must not be allowed to disrupt them in any way. If a robot is beneficial for customers but a major nuisance for the staff, then we don't see it as a success.

**HOW COULD THE** ROBOT **BE A VALUABLE ASSET FOR THE STAFF AS WELL?**

J.L.    I believe a robot will show its strengths where humans have their limits. Among other things, a robot is always responsive, aside from issues with battery life and other such factors. A robot could, say, take over tasks at night, which would help

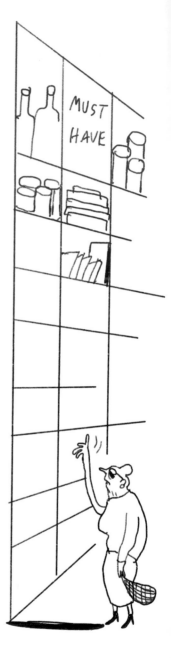

to streamline the work routine before the store opens in the morning, thereby reducing the workload of the employees. Employees have various roles. They can order, accept and re-stock merchandise, help customers, handle the checkout, and are experts in product knowledge. It's challenging when your responsibilities consist of five minutes at the checkout, then three minutes of restocking, and then checkout again for five minutes. The technology is already set up for that, like express checkout lanes, which can be opened and closed at short notice. For the employee, constant changes can be stressful. Our robot could assist in keeping the workflow more consistent, which has been proven to help make the workday more pleasant.

**WHAT HAS TO HAPPEN IN SOCIETY FOR SHOPPING ROBOTS TO BECOME ESTABLISHED THROUGHOUT GERMANY? DOES THE SUPERMARKET PERHAPS NEED TO CHANGE TOO AND NOT JUST THE ROBOTS?**

J.L.    Many processes that are hidden from customers are already supported and optimized by the use of robotics. For the staff, this is a huge step forward and a relief. This applies, for example, to the efficient selection of goods in the warehouse by robots and the resulting high availability and quality of fresh products in the supermarket. I would not say that average customers explicitly want a robot for the robot's sake. They are simply looking for a special shopping experience.

G.B.    It is not the technological deficits that stand in the way of the widespread use of robots in supermarkets. Our project shows that the required technology already exists. It needs to be improved, miniaturized and to become invisible. A larger problem is the missing legislative framework for the use of autonomously acting robots in public spaces. It must be clear what happens if, for example, despite all precautions, an accident occurs involving an autonomously acting robot.

**WHAT OTHER OPPORTUNITIES AND CHALLENGES HAVE EMERGED IN THE PROJECT?**

G.B.    I was surprised to see how much technology is already implemented in today's supermarkets and how systematically technological development is being pursued, for example, in the field of digitalization. There are already well-defined interfaces, for example, for the digital registration of goods that can be used to directly integrate new technology such as robots.

**J.L.** Yes, it usually seems unremarkable, but the underlying processes are very complex. Yet they offer a wide range of possibilities for sourcing information on products, shelf placement and other basic and movement data usable for the robot. We were impressed by the challenge that a supermarket shelf poses to a robot; the complexity of planning a secure retrieval of an item surprised us. I believe that there is still enormous untapped potential to assign other tasks to the robot as well. Some examples could be moving products to the front of the shelve and checking expiration dates or putting away new merchandise.

**A.B.** The challenges lie in the *soft* topics, for example, data protection. A robot needs comprehensive sensing technologies in order to recognize things. Of course, this also includes cameras, but what are the consequences for privacy? Such legal and social aspects must be taken into account in the development of robots.

**G.B.** For further development, we have to get robots out of the lab! It is a different matter to program a robot to stock shelves in a lab than in a public space. Whenever humans are involved, a new set of requirements has to be taken into account. For example, the robot can try to make personalized offers to customers based on the recognition of intentions and specific circumstances. Doing this, however, the robot may come so close to the humans that they feel uncomfortable. Another example involves elderly people or people with special needs. While the robot is certainly able to identify such customers, it is a more difficult problem to decide whether it is appropriate for the robot to offer help. Such questions actually have little to do with what developers are usually concerned with in a lab. Such requirements can only be defined in *real-world* projects.

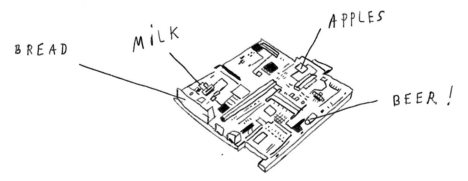

Inspired by the interview, the illustrator Johanna Benz visually commented on chances, risks and scenarios of robotics from her own perspective. © Johanna Benz

# RobotKoop

### AIM OF OUR RESEARCH PROJECT

We develop cooperative interaction strategies for service robots in both private and public spaces. The idea is to create robots as friendly, communicative and competent interaction partners that understand the user, adapt to social circumstances and other environments, and learn new tasks.

### CONTEXT, ROLE AND TASK OF OUR ROBOTS

There are two service robots in two application domains. One is a cleaning robot for public spaces that cleans the floors and interacts with passers-by when necessary (for example, adaptation of driving path). The other is a humanoid domestic robot that helps with daily chores and is a cooperative, adaptable, and teachable assistant.

### WHAT MEANINGFUL HUMAN-ROBOT INTERACTION MEANS TO US

Human-robot interaction should be intuitive, acceptable, trustworthy and safe in the public and domestic context. Robots should be teachable and able to adapt to the user rather than dealing with all users in the same way.

### WHO WE ARE

University of Ulm, Institute of Psychology and Education, The Human Factors Department; Ulm University, Institute of Communications Engineering; ADLATUS Robotics GmbH; InMach Intelligente Maschinen GmbH; University of Applied Sciences Ravensburg-Weingarten, The Institute of Artificial Intelligence, IKI.

# Dominant, Persuasive or Polite?— Human Curiosity, Provocative Users and Solving Conflicts between Humans and Robots

*Interview with*

Franziska Babel (F.B.)
Siegfied Hochdorfer (S.H.)

*By*

Judith Dörrenbächer

**YOUR CLEANING ROBOT FOR PUBLIC SPACES ENCOUNTERS VERY DIFFERENT PEOPLE EVERY DAY WHO ARE NOT THE ROBOT'S OWNERS. WHAT CHALLENGES DID YOU IDENTIFY?**

**F.B.**    Unlike at home, people in public spaces are not prepared to face a robot. Concerning the interaction design, we therefore had to consider what would be an acceptable spatial distance between robots and humans. Another question was how the robot should deal with disturbances, that is, what if humans get in its way? Our observational study, however, took a surprising turn against our expectations: people are already using autonomous technology very naturally. We thought we would encounter more problems.

**DO YOU HAVE A HYPOTHESIS AS TO WHY THAT IS?**

F.B.     I believe science fiction has normalized the idea of autonomous vehicles. Some people therefore even overestimate
the technology.

**DOES THAT MEAN YOU HAVE TO ADAPT YOUR** ROBOTS **TO
THESE SCIENCE FICTION VISIONS TO PREVENT MISUNDER-
STANDINGS?**

F.B.     I don't think it's necessary to copy what is shown in the
movies, but one should keep in mind—when designing—that
people have seen them and expect certain behaviors. Otherwise, they *over-trust* the robot and accidents might happen.
Ultimately, one has to make transparent what a robot can
actually do before it's applied, for example, through information campaigns.

S.H.     Yes, the robot should be deliberately designed to meet its
capabilities. For example, designers must not overdo anthropomorphism. High expectations are easily disappointed
when you realize: "It keeps repeating the same phrases while
standing on that same corner every single day." But our
project was also about safety. Of course, robots have to be
designed in such a way that accidents definitely won't
happen. In this respect, much higher standards apply in public spaces than in industrial environments, because here
the robot also encounters small children or pets.

HE is SO
ANTHROPOMORPH...

**HAVE YOU ALSO DEALT WITH THE ISSUE OF** ROBOT **BULLYING IN THIS CONTEXT?** ROBOTS **IN PUBLIC SPACES ARE OFTEN THE** <u>VICTIMS</u> **OF VANDALISM.**

**S.H.**  Yes, but we have hardly seen any vandalism. This is a very rare phenomenon with the many robots we have in use. People are more likely to start playing around with the robot or push its behavior to the limit in order to find out what works and what doesn't. When a robot is active for ten minutes, you see it at least once. You don't have to wait for long; it starts right away: The few existing buttons are pushed and people try, for example, to stop the robot. And if it stops, some people try to get it going again.

**WHAT MAKES THE** ROBOT **SO INTERESTING FOR THESE INTERVENTIONS COMPARED TO OTHER OBJECTS IN PUBLIC SPACE, SUCH AS JUNCTION BOXES OR PARK BENCHES?**

**F.B.**  Well, with a robot, it certainly plays a role in that it moves, swerves and stops. You want to know—perhaps out of curiosity—how it reacts. Does it fight back at some point? Is there any level of escalation that I haven't seen yet? A park bench won't react, but a robot might. Most people probably believe the robot will ultimately not fight back. Unlike a stray dog that eventually bites, they think: "The robot is programmed according to certain security guidelines not to hit me no matter what I do."

**YOU HAVE INTENSIVELY DEALT WITH THE TOPIC OF ACCEPTANCE. WHAT DID YOU FIND OUT?**

**F.B.**  We conducted two observational studies and asked participants: "What do you think of this robot? Do you have any concerns about this robot operating here or at home?" We were also interested in what a robot is, and is not, allowed to do. The topic of privacy arose quickly in the domestic context. The cameras of our robot *TIAGO* were often interpreted as eyes and thus the suspicion arose that the robot was monitoring its surroundings. Interestingly, compared to smartphones—which most people own and which also have cameras—surveillance comes to people's minds much quicker. With *TIAGO*, people also remarked: "I don't have space to put it in my apartment," or "it can't climb stairs." We wouldn't have noticed these everyday problems without our study. With the cleaning robot, we saw that speech might not be the best means of communication because the robot doesn't look human—beeps or movements might be more appropriate. In contrast, with the home robot, people were more confused when it didn't speak. In the public area, people said, "nah, it should just do its job and not talk to me," while in the home area, the robot should ask before it throws something in the trash. I would recommend all future projects to conduct such observational studies early on.

**DO YOU THINK THAT ANTHROPOMORPHIZING THE ROBOT PLAYS A BIG ROLE IN ACCEPTANCE?**

F.B.    Yes, and apparently not only in a positive sense. I think a robot design without eyes would make people feel less observed. Maybe it even makes sense to simply not install cameras if they are not necessary, as with the cleaning robot in our project.

TRASH

TRASH

**YOU ALSO LOOKED AT THE ISSUE OF CONFLICTS OF INTEREST. WHAT IS MEANT BY THAT? WHY DO CONFLICTS ARISE BETWEEN HUMANS AND ROBOTS?**

F.B.    We are mainly interested in conflicting goals between robots and humans. For example, the robot wants to clean in front of the lockers at the train station, but the human wants to put suitcases into said lockers. These goals cannot be achieved at the same time—a conflict arises.

**ARE ROBOTS DIFFERENT FROM OTHER TECHNOLOGY IN THIS RESPECT?**

F.B.    Yes, because the robot operates autonomously and has its own tasks and goals. Immobile tools, like a manual lawn mower, will not create conflict on their own. However, there is not only a difference with other technologies, but also with humans. With another human being, I accept that they could ask me to get out of the way because the other person has exactly the same right to exist as I do. Robots, on the other hand, are considered to be subordinate machines and are therefore often expected to wait. We assume—that's why we might have more conflicts between humans and robots than between humans and humans in the public context.

NO!

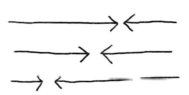

### HAVE YOU DEVELOPED INTERACTION STRATEGIES FOR RESOLVING THESE CONFLICTS?

**F.B.**  Yes, we tested different robot interaction strategies, from "very polite and persuasive"—here the robot convinces by presenting an advantage or explaining its goals—to "very assertive"—here the robot appears very determined, gives orders, or approaches closely to show: "I want to go right here now." We noticed that polite strategies are more acceptable but not always effective, while a commanding robot can be more effective but sometimes elicits more negative reactions; we had to find a balance between politeness and assertiveness. Persuading the human with arguments has worked relatively well. The robot might say something like: "If you step aside for a minute, I can continue cleaning here; it will not take much longer."

### WHAT DID YOU EXPERIENCE IN THE DOMESTIC SETTING?

**F.B.**  At home, an adaptive strategy could be more suitable. Designers could make use of questionnaires to find out whether users have a dominant or passive personality and let the robot adapt accordingly. Alternatively, the robot could ask something straightforward, like: "How shall I ask you to get out of the way?" In fact, we found that in the home, the power imbalance is significantly greater. After all, in the public space, I might think: "Well, this is a representative of the cleaning company. Similar to a human cleaner, I'll get out of its way." The robot has a certain authority there. At home, I have bought the robot myself. I own the robot. There, the idea is: "It should do what I want it to do. I should not be the one receiving orders."

### IN THE FUTURE, WHAT KIND OF ROLES DO YOU THINK ROBOTS COULD PLAY IN OUR SOCIETY? WHAT KIND OF FUTURE ARE WE HEADED TOWARDS?

**S.H.**  I think robots will be everywhere. Being a robotics researcher, I'm a bit more optimistic here. In Germany, however, regulations are an obstacle. We have a very good and strong data protection policy, which is an obstacle from a purely technical point of view. In many Asian countries, in contrast, robots are allowed to move around in public spaces with camera systems. The safety requirements for robots are so high in Germany that there are currently not even safety sensors for public spaces that cover the standards. All safety workshops we participated in ended with the conclusion that there is no <u>solution for public spaces</u>. Of course, the entrepreneurial risk is very high because of that, and other countries will proceed much faster.

F.B.     I think people will increasingly realize that robots don't take away jobs from humans. Especially during the Covid-19 pandemic, robots could have helped a lot in nursing, because they don't transmit diseases.

S.H.     Yes, we have a labor shortage in many areas. Many companies asking us for cleaning robots are not concerned with cost efficiency, but quite simply with the fact that no one wants to do this kind of work. Robots also create a lot of new jobs since robots can't look after themselves. A lot of service jobs are being created that we don't even know about yet. It is comparable to the change from horse-drawn carriages to automobiles. This change didn't lead to fewer jobs, but to more, despite the fact that a car can transport things much faster. Correspondingly, I don't think robots will ever completely replace people.

F.B.     Yes, there will always be a coexistence of humans and robots. Both have advantages and disadvantages and can support each other and be useful in a variety of ways.

DOMINANT, PERSUASIVE OR POLITE?

FRANZISKA BABEL, SIEGFIED HOCHDORFER

Inspired by the interview, the illustrator Johanna Benz visually commented on chances, risks and scenarios of robotics from her own perspective. © Johanna Benz

# James Auger

is an Enseignant Chercheur and Directeur of the department of design at the *École Normale Supérieure Paris-Saclay.* His practice-based design research examines the social, cultural and personal impacts of technology and the products that exist as a result of its development and application.

# Seven Observations, or Why Domestic Robots are Struggling to Enter the Habitats of Everyday Life

James Auger

*Close your eyes, empty your mind ... Now, picture a* robot.

The robot holds a unique place in Western culture—heavily shaped by a long history of fictional and spectacular representations of artificial life. For the past century, robots have also become symbols of technological futures. These typically fit neatly into two polemical opposites:

**UTOPIAN:**

Following modernist notions of progress, robots (typically in corporate and research culture) are used to represent an advanced future: according to Langdon Winner, "it is not uncommon for the advent of a new technology to provide for flights of utopian fancy" (1986, p. 106). The robot has taken us on many such journeys: from the metallic humanoids of the 1930s, such as *Westinghouse's Electro*; the promise of the cyborg in the late 1990s, such as the academic research of Kevin Warwick; to the miniature medical nano-bots being proposed today.

**DYSTOPIAN:**

Cautionary tales related to the creation of life, such as the Jewish legend of the Golem, Greek mythologies such as *Pygmalion* and *Daedalus*, and Mary Shelly's *Frankenstein*, have for millennia warned of the dark consequences of humans playing God. Karel Čapek, in his 1920 play *R.U.R. (Rossum's Universal Robots)*, continued this theme through the story of a factory that makes artificial human workers from synthetic organic matter (at the same time coining the term robot). These fictitious works disseminate powerful and memorable negative constructions into the public domain, tainting perceptions, tarnishing the robot's reputation and constructing a shared negative imaginary.

In the years following *R.U.R.*, fiction and reality diverged as automation came true, but production-line robot design emerged as a logical progression of Fordism, and as such the pragmatic demands of manufacturing engineering informed their form and behavior (rather than the notion of the humanoid). The success of industrial robots, however, has done little to diminish our fascination with the more sensational robots from fiction, and it is these spectacular versions (typically existing as anthropomorphic, zoomorphic or paedomorphic devices) that continue to dominate robotic research and development (and most likely the depiction I placed in your mind's eye at the beginning of this text).

 Such *spectacular* robots are thriving in research labs, technology fairs and science fiction. In these habitats a robot's raison d'être is justified by fulfilling requirements very different to those of everyday life. The promise, however, is that one day soon these robots will move seamlessly into the domestic domain. It is the potential of this transition that feeds many of the dreams of robotic futures, and has done so for almost a century. The robot in domestic life has become a recurring dream, either because those responsible for these promises have no interest in realising the dream, or, more likely, because they drastically underestimate the complexity of the journey—most robots would need to go through a process of huge adaptation to achieve this goal.

 To assist in this process, a subtle change in perspective could be helpful—viewing the robot as a product rather than a symbolic icon would introduce design into the equation. Design, by its very nature, is about everyday life. It is through the intervention of the designer that technological potential is transformed into the usable and useful products of daily life, through the crafting of elements such as form, function and interaction. Designers also address more complex human issues, such as the desires and nuances of an intended audience, the role of fashion and trends, and more pragmatic issues such as cost vs. benefit. By exploring the conditions for a successful existence in the domestic space, it becomes possible to offer some answers to the obvious disparity between the promise of the spectacular robot, as imagined by film directors, writers, scientists and corporations, and what is actually available to the public today.

JAMES AUGER

I will conclude with a few key statements on why robots have yet to be domesticated.

## 1. MY HOME IS NOT A LABORATORY

In his influential book *The Ecological Approach to Visual Perception*, James J. Gibson stresses the value of moving out of the laboratory (1986, p. 7). If robots are to enter our homes at some future point, roboticists would do well to heed Gibson's advice: "The laboratory must be lifelike!" (p. 3)—only then will the combined value of their form and function, and the way we interact with them, provide the necessary incentives for living with them.

## 2. THE ROBOT IS TOO OFTEN A SOLUTION IN SEARCH OF A PROBLEM

Successful products are usually conceived in response to a problem. Fictional notions have produced robots that can walk, shake hands and even communicate with humans. The fundamental question that should be asked at the outset is: why is this necessary? To borrow from Cedric Price, if a robot is the answer, what was the question?

## 3. OR, THE PROBLEM IS THE PROBLEM

The problems resolved through robot research are commonly related to the (very difficult) challenges of replicating forms of natural life. These can be motivated by the challenge itself rather than a justification of why it is necessary. The massive problems of artificial intelligence, or even creating a machine that can walk up a flight of stairs, become intellectual pleasures that consume all focus. This has the consequence of rendering unimportant all questions related to applications, implications, reality …

## 4. THE EVOLUTIONARY NATURE OF PRODUCTS

Technological artifacts typically follow predictable evolutionary iterative pathways. Robots in their current guises have no such lineage in the contexts of everyday life; rather, their lineage exists in fiction and storytelling. This raises the question of how robots evolve. What are they currently being adapted to, and how might a lineage start in the domestic habitat?

## 5. THE ROBOT PARADOX #1—SUBLIMITY AND EPHEMERALITY

Robots have a special status above that of normal products. The problem arises when this special or *sublime* element is provided by novelty and curiosity (for example, *Sony's Aibo*). For, as Edmund Burke points out, "curiosity is the most superficial of all the affections; it changes its object perpetually" (2008, p. 29). Robots have effectively become victims of their success as vehicles used to represent the future, through the use of technological novelty to provide spectacle. This will always be an ephemeral success.

### 6. THE ROBOT PARADOX #2—HEIMLICH AND UNHEIMLICH

The home: bastion of comfort, security and the familiar. The robot: alien, mysterious and threatening. These opposing traits need to converge. The few robots that have successfully entered the home follow the forms of traditional products. For example, robot vacuum cleaners and lawn mowers are effectively roboticised versions of existing objects. In stark contrast, *spectacular* robots would represent the introduction of a very foreign species into the home—this would be analogous to inviting a wolf in through the door.

### 7. PLAYING TO STRENGTHS—IMPLICATIONS OF A (NON-FICTIONAL) ROBOTIC FUTURE

As robot-related technologies become more sophisticated, their ability to gather increasingly sensitive or emotive information becomes a genuine possibility; this, combined with the computer's ability to store and analyze data, could result in a future saturated with automation. Sensitive and emotive tasks become delegated to the device as trust grows (*Amazon's Echo* is like a Trojan horse in this respect). When robotic technology is employed to play to its strengths, the potential consequence is that we in turn become robotic—less emotional, more rational, programmed and predictable; technology effectively replaces human thought. We need to be very careful with what we choose to automate.

Burke, E. (2008). *A philosophical enquiry*. Oxford University Press.

Gibson, J.J. (1986). *The ecological approach to visual perception*. Psychology Press.

Winner, L. (1986). *The whale and the reactor*. The University of Chicago Press.

PERSPECTIVES

SEVEN OBSERVATIONS, OR WHY DOMESTIC ROBOTS ARE STRUGGLING
TO ENTER THE HABITATS OF EVERYDAY LIFE

JAMES AUGER

# INTUITIV

### AIM OF OUR RESEARCH PROJECT

We aim to design hospital robots that behave in an intuitive manner, are non-detrimental to the sense of wellbeing of all present stakeholders, and support staff and patients in everyday tasks and activities.

### CONTEXT, ROLE AND TASK OF OUR ROBOTS

We work with three different robotic platforms: two mobile robots—a robotic rollator and a transport robot—accompany and support patients at rehabilitation clinics in various situations, such as guiding patients to their appointments or rooms while also transporting luggage. Another robot is a stationary robot arm that supports the clinic staff with handling medical equipment.

### MOST ASTONISHING FINDING LEARNED ABOUT ROBOTS DURING OUR WORK

Many participants in our studies did not wish for the robots to have an anthropomorphic or zoomorphic look, realistic avatars, or an elevated disposition towards social interactions—"these are clearly tools—albeit smart ones—so they should act and look like that as well."

### WHO WE ARE

ek robotics GmbH; HFC Human-Factors-Consult GmbH; Rehabilitation clinic Johannesbad Fachklinik, Gesund-heits- & Rehazentrum Saarschleife; DFKI GmbH, The German Research Center for Artificial Intelligence, Cyber-Physical Systems CPS in Bremen and Cognitive Assistance Systems COS and Multilingual Technologies MLT in Saarbrücken; Gestalt Robotics GmbH.

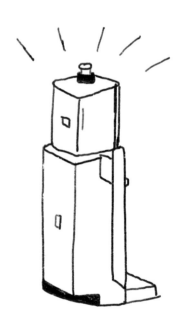

# Is this a Patient or a Wall?— Adapting Robots from an Industrial Context to a Rehabilitation Clinic

*Interview with*

**By**

Karsten Bohlmann (K.B.)
Tim Schwartz (T.S.)
Hanns-Peter Horn (H.H.)

Jochen Feitsch
Bernhard Weber

**WHAT WILL THE ROBOTS DEVELOPED IN YOUR INTUITIV PROJECT BE USED FOR AND WHAT ARE THEIR MAIN FEATURES?**

**K.B.**   In the case of our transport robot, the end users are the staff in rehabilitation clinics that are spread over several interconected buildings. There are long distances and it is easy to get lost. To support the clinic staff, robots are to take over tasks such as transporting luggage, assisting with walking, guidance throughout the buildings and providing information to the patients.

**COULD THE ROBOT ALSO POINT OUT IMPORTANT LOCATIONS ALONG THE WAY, SUCH AS THE CAFETERIA?**

**K.B.** Yes, certain points of interest are defined, and when the robot passes these, it can start a corresponding dialogue with the patient.

**WOULD A USE BEYOND THE CLINICAL AREA ALSO BE CONCEIVABLE, FOR EXAMPLE, IN HOTELS OR AIRPORTS?**

**K.B.** Definitely. Once the technology is available, other areas will increasingly come into consideration; there will be significantly more use cases in the future, especially for accompanying people and transporting objects.

**WHICH ROBOTS ARE DEVELOPED IN INTUITIV BESIDES THE TRANSPORT ROBOT?**

**T.S.** There is also the autonomous rollator (or wheeled walker) and the *Franka* robot. The latter is a one-armed robot that is supposed to help with blood collection by, for example, passing, lifting or lowering a tray with the needed tools on command. We were able to implement this technically, but the professional clinic staff was already working very fast. Thus, the robot could not provide them with any real support. The situation is different with the autonomous rollator, because older patients are often cognitively overwhelmed by finding the right clinic rooms. The rollator can be requested and drives up to them autonomously. From then on, it must be pushed normally, but can guide the patients to their destination with route information.

**WHAT HAVE BEEN THE BIGGEST CHALLENGES IN THE PROJECT OVER THE YEARS?**

K.B.    Fluid robotic movements through groups of people with compliance to safety requirements and the redesign of the robot's construction were especially challenging. We revised an existing robot technology intended for industry use and redesigned it to become an assistance robot for public areas, especially designed for interacting with people. Unlike in an industry setting, the robot has to navigate in less structured environments and must therefore detect and avoid small or overhanging obstacles. An additional challenge is the processing of 3D and visual sensor data. Here, a lot of onboard computing capacity is required, which we must provide for the robot. In rehabilitation clinics, we cannot assume that wi-fi with sufficient bandwidth will be available everywhere.

**CAN THE ROBOT DISTINGUISH STATIC OBJECTS FROM MOVING PEOPLE?**

K.B.    Making this possible is one of the current tasks, especially for our partners *Gestalt Robotics* and *DFKI.* In principle, we need real-time human detection.

T.S.    A lot of work is now being done with neural networks. However, conventional computers are not yet necessarily prepared for AI applications; they lack the necessary graphics power. For the detection of humans in real-time, *Gestalt Robotics* has developed a module that uses depth camera data. For avatar control, we combine this data with that from facial detection. The depth camera component says, "I think there's a human," we check to see if there is a face, and then we can increase the confidence that it's a human. In addition, we use microphone arrays to determine the direction of speech signals.

**HOW DO YOU ASSESS THE OPPORTUNITIES AND POSSIBILITIES OF VIRTUAL REALITY (VR) TECHNOLOGY FOR HUMAN-ROBOT INTERACTION?**

H.H.    Thanks to VR, we were already able to gain initial insights into the interaction between robots and humans before we were able to test with physical hardware. A virtual robot is generated, and the user is then enabled to interact with this virtual robot. Ideally, the user will have the impression that they are fully immersed into the virtual scene and interact with the robot like in a real-world situation. Without VR, testing like this usually starts later, and as a result, there is often not enough time and resources left to start over again if there are problems. VR could also be helpful to put oneself (or the user) *into the shoes* of the virtual robot to get a better impression of the capabilities and remaining limitations during development.

**YOU MENTIONED THAT IMMERSION IS AN IMPORTANT FACTOR. HOW MUCH COULD THE USERS IMMERSE THEMSELVES INTO THE FICTIONAL SCENE OF YOUR VR STUDIES?**

H.H.  We conducted several VR studies to investigate which of the robot's motion strategies are perceived as convenient. Although our simulation did not yet look overly realistic, the immersion was surprisingly very strong. This was not only revealed by the questionnaires, but there were repeated amusing moments where study participants indeed tried to stand on top of the virtual robot's loading space to drive around on it.

**WHY WAS A LOT OF EMPHASIS PLACED ON NON-VERBAL COMMUNICATION IN YOUR PROJECT?**

T.S.  Because constant speech interaction can be annoying, *DFKI* developed a rule-based dialogue system that uses environmental and contextual knowledge to produce dialogue acts. Humans also use facial expressions, gestures, and their eyes for communication, such as when we want to pass by each other. We wanted the robot to be able to communicate in a similarly intuitive way. Thus, the robot also uses other indicators for this purpose and uses its eyes to give the impression that it is looking at its counterpart and is reacting to their movements.

H.H.  It's all about creating an efficient coexistence and avoiding discomfort. When robots and humans have to pass each other in a narrow hallway, we achieve this by letting the robot start its evasive maneuver as early as possible and using its turn signals. This is a logic that everyone knows and understands.

**DOES THE ROBOT KNOW WHERE THE HUMAN WANTS TO GO?**

T.S.  It can make an estimate based on the sensor data, but we can't completely rely on intention recognition. It's safer for humans if the robot indicates what it's going to do.

**K.B.** The robot is equipped with safety scanners that build up so-called *protective fields,* which is a mandatory safety feature. The moment any obstacle enters the protective field, the robot stops immediately. To avoid continuously repeating this behavior, the robot constantly adapts its path and speed to the people in its environment, like when following a person. However, a robot in public spaces should also display a certain assertiveness. It does not make sense for it to constantly stop or change sides of the walkway when it detects a person three meters away.

**H.H.** In the second study, we were fascinated to find that about half of our study participants expected the robot to submit at all times. The other half perceived it as an interaction partner, that is, wanted to let it through, and reacted irritated when it did not acknowledge their forthcoming offer. Hence, a permanently submissive robot is not necessarily the best solution.

**WHICH TECHNICAL DEVELOPMENTS ARE STILL MISSING FOR EVERYDAY USE OF THE** ROBOTS**?**

**K.B.** We need to validate our concepts in field tests. We have to find out what is still lacking and then improve the system accordingly. However, I believe that we are not considerably far away from suitability for everyday use.

**T.S.** There is still a long way to go, but a robot with a limited range of functions can be implemented in a relatively stable way. For communication, however, it is important to have a large language corpus and to understand what people expect from the robots and how they express it. Something like this takes time and, despite machine learning, still a lot of manual work.

**WHAT HAS TO HAPPEN IN SOCIETY FOR ASSISTANCE ROBOTICS TO BECOME ESTABLISHED?**

**H.H.** Many people think of *Robocop* or *Wall-E* when they think of robots. Such visions are still unrealistic, even at today's level. In addition, robots are expensive and will probably remain so for a long time. There would have to be more acceptance among the population that most of them will be useful helpers, but not necessarily overly intelligent.

**K.B.** I think, in the future, people will perceive robots as everyday technical devices—think of robotic vacuum cleaners—and there will be a gradual development towards more robotic helpers in society. People will expect a minimum intelligence from robotic assistants, but robots won't be perceived as persons. It also depends on which society we're talking about. For example, parts of Asia and the US seem to be quite open to the idea of robots in everyday life.

**T.S.** I would rather turn the question around: What do the developers of robots have to do so that assistance robots will be accepted in society? Robots have to adapt to society and not the other way around, and we, as developers, are responsible for that.

**ARE THERE ANY OTHER INSIGHTS FROM THE PROJECT THAT YOU WOULD LIKE TO SHARE WITH US? WHAT CHALLENGES DOES ROBOTICS IN GENERAL STILL FACE?**

**H.H.** It will become increasingly important for us to sharpen our focus on the range of functions a robot should have. In the robotics community, I sometimes notice that people are looking for the perfect all-in-one solution. Oftentimes a simple, less anthropomorphic design can already communicate a lot of the limitations that have to be expected. The more features that are added, the more complex it becomes and the greater the chances that something will go wrong.

ALL-IN-ONE, PLEASE

ROBOTS

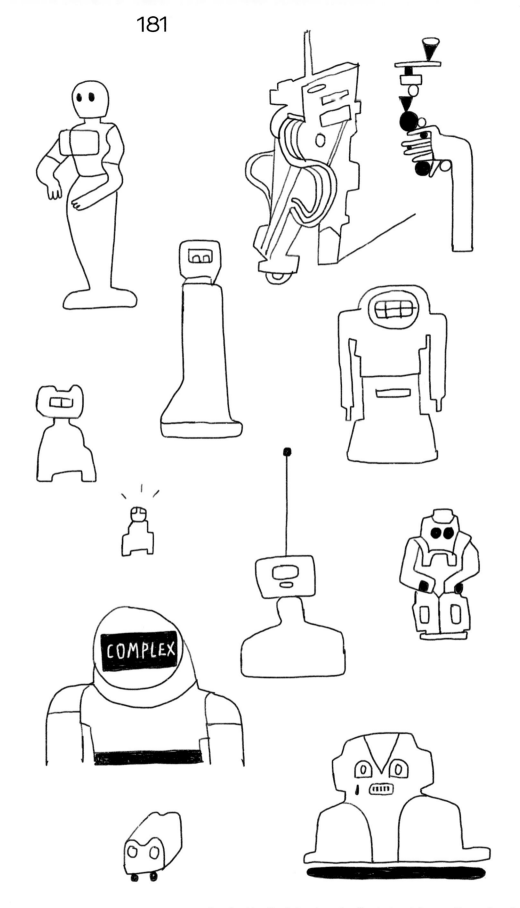

COMPLEX

Inspired by the interview, the illustrator Johanna Benz visually commented on chances, risks and scenarios of robotics from her own perspective. © Johanna Benz

# Dr. Corinna Norrick-Rühl

is Professor of Book Studies in the *English Department at the University of Muenster*, Germany. In her research and teaching, she focuses on 20th- and 21st-century publishing history and book culture. Recent publications include *Book Clubs and Book Commerce* (Cambridge University Press, 2020) and the volume *The Novel as Network: Forms, Ideas, Commodities* (Palgrave, 2020, co-edited with T. Lanzendörfer).

# Robotics x Book Studies—Imagining a Robotic Archive of Embodied Knowledge

Corinna Norrick-Rühl

Imagine you are visiting a museum of technology in 2052. The machines and tools are there, but they are not gathering dust. It is loud, and there is an abundance of smells—some quite pungent—in the air. The clanking of machines and the hustle-bustle of production fills the domed hall. You almost feel overwhelmed as you pass large 19th-century power looms weaving sumptuous cloths and continue past a loud area in which rag-based paper is being made, with hammers pulping linen rags. You hurry onwards to a quieter, more subdued area of the hall, where movable type is being set, methodically, at a big wooden type case. When you get close, you can hear the efficient click-clack as the type is placed into the composing stick. A few steps further away, the air is heavy with a sweet inky smell, and texts are being printed, just as they would have been before the advent of the Linotype line-casting machine, photocomposition and desktop publishing.

You are experiencing a *living museum*, not only documenting the past in immobile exhibits, through videos or informative panels, but also creating and re-creating the objects that have shaped human history. Or maybe it isn't *living*—the operators, at least, don't seem to be alive? The operators, in fact, are not human, though they move about like humans and bear resemblance to the human form. And in this museum, against all odds, they are performing the "heritage skill" of letterpress printing (Maid in Britain, 2018) with surprising dexterity and purpose.

As a society, we had lost touch with the arts and crafts which shaped our history. Simply put, there were not enough humans left who were experts in the historic crafts and who could act as *interpreters* of the past in a museum, or who could reliably teach future generations these crafts, which no longer had an industrial use. By the 2030s, the situation was urgent. To prevent permanent loss of these technologies, and of the embodied cultural knowledge that is needed to work with them, teams trained anthropomorphic robots to act as animate archives, simultaneously able to offer training to potential (human) apprentices and to safely conserve the knowledge when no potential apprentices are waiting in the wings. The interdisciplinary team of (among others) cultural historians, robotics experts and mechanical engineers joined forces to form a robotic archive of embodied knowledge in the 2030s and 2040s. All around the globe, in museums, the results of this innovative teamwork can be viewed, and the rewards can be reaped.

For now, in the year 2022, this is merely a thought experiment, and is certainly not feasible (or fundable!) within a realistic timeframe. For now, documentary videos and, perhaps, virtual reality options are our best bets. However, as a book historian, I am deeply concerned about the loss of knowledge about the arts and crafts of papermaking and printing, from typecasting to typesetting and letterpress printing. There is an urgent need for knowledge transfer to protect the cultural heritage of these and other papermaking and printing techniques. In 2012, the documentary *Linotype: The Film* drove this point home in its trailer. "How does the Linotype fit in with new technology?", the interviewer asks behind the camera. Without missing a beat, the printer answers, "It doesn't" (*Linotype: The Film* [Trailer], 2012, 01.46–01.50). Nonetheless, the Linotype shaped newspaper history—and is thus closely entangled with our collective cultural history.
  Recently, the *Mainz Impulse* (Gutenburg-Museum, 2020) drew attention to the fact that museums around the world are struggling to find knowledgeable individuals, either as employees or volunteers, to conserve and pass on the processes, step-by-step, movement-by-movement. It seems that the machines will survive in printing museums and museums of technology, but the knowledgeable human touch to operate them is dying out. Worldwide, only very few individuals still possess the know-how and experience to set type by hand, or to use a Linotype machine. Perhaps we could try to think outside the box, and embrace those 21st-century technologies that, somewhat ironically, made older technologies redundant to begin with? Which leads me, again, to wonder: what if anthropomorphic robots could help us conserve our cultural heritage, starting with an archive of embodied printmaking knowledge? Can new robotic technologies in fact help conserve and protect older technologies, or

is this problematic and perhaps even unethical? Is preserving the old with the new merely a fallacy, a form of *botoxing*, and thereby weakening, even paralyzing the power of our cultural heritage? Staying with this metaphor, at first glance the advanced technology seems to be a reliable measure of conservation, but ultimately it robs the arts and crafts of their authenticity and opportunity for expression.

As James Fallows recently wrote in *The New York Times*, "The most persistent and touching error [of humankind] has been the ever-dashed hope that, as machines are able to do more work, human beings will be freed to do less, and will have more time for culture and contemplation" (2021). Certainly, we shouldn't rely on robotics to take care of culture and contemplation for us. But perhaps we can think productively and creatively about engaging robotics to protect and conserve the arts and crafts of culture and contemplation.

Fallows, J. (2021). Can humans be replaced by machines? *The New York Times*. Retrieved May 4, 2021, from https://www.nytimes.com/2021/03/19/books/review/genius-makers-cade-metz-futureproof-kevin-roose.html

Gutenburg-Museum (2020). Mainzer Impuls. Retrieved May 4, 2021, from https://www.mainz.de/microsite/gutenberg-museum/global/2020_mainzer_impuls.php#c6

*Linotype: The Film* (Trailer) (2012). YouTube. Retrieved May 4, 2021, from https://www.youtube.com/watch?v=avDuKuBNuCk

Maid in Britain (2018, April 19). The sensory pleasures of printing and pressing. Retrieved May 4, 2021, from https://maidinbritain.com/2018/04/19/the-sensory-pleasures-of-printing-and-pressing/

# Antje Herden

has written popular books for children and young adults for publishing houses like *Tulipan, Fischer*, and *Beltz*. She was awarded the Peter Härtling Prize in 2019 for *Keine halbe Sachen*, and was nominated for the German Children's Literature Award. She lives in Darmstadt and has two children.

# "That's the Future, I'm Telling You."

*A conversation overheard by Antje Herden*

"There are robots in the supermarket that go shopping for you now. That's the future, I'm telling you. And we're already there. Incredible. Absolutely incredible. Cheers, darling."

"What do you mean, they go shopping for you? Do you send them out with a list and then they bring everything to your doorstep?"

"No. They're out in front of the store. They look like shopping carts. You tell one of them what you want, then it rolls through the aisles and picks out everything you requested."

"I see. But you still have to go to the supermarket, right? And what do you do then? Just wait there until the robot is finished? Do you walk alongside it while it picks everything out? Sounds kinda silly."

"You can go look for other things in the meantime. I, for instance, like to look around in the fruit department while the robot goes and gets the things that are boring."

"The things that are boring? Scientists are wasting loads of resources to develop robots because shopping can be boring?"

"It's just that I don't like to hang around the hygiene articles, that's all. Tampons, condoms, the whole shebang. We need new lube too, but I don't feel comfortable looking for it on the shelves."

"Why? Because somebody might be disgusted that you have sex? At your age? Ha."

"It's not funny."

"Sorry. So, the robot got you the lube because you were too uncomfortable to."

"Well, not really. It asked what I needed, and I told it. But it didn't understand me."

"Isn't lube in its program?"

"I said it too quietly. I didn't want to shout through the whole supermarket that Marcus and I needed some lube. So I went and found it myself. Meanwhile, the robot was in the fruit department."

"In the fruit department? I thought—"

"I didn't dare go to the fruit. There were too many robots there. I was afraid they'd run me over. So I just ordered some apples. After all, the fruit isn't always ripe. You have to take a look at it before making a decision."

"The market."

"Sorry?"

"There's a weekly market on Thursdays. It's great."

"You're pulling my leg."

"Just a little."

"But seriously, a robot like this is really helpful for people with disabilities. You can't reach everything in a wheelchair."

"But you could if the supermarkets were made more accessible for the disabled. Besides, human communication still exists. Some people go shopping for precisely that. To talk with each other. To ask for help sometimes, too. I'm always happy to help."

"But sometimes there's no one there in the vast expanses of the aisles."

"Why do we even need such endless expanses of aisles lined with shelves? Nobody can eat all of these groceries. Everyone's overwhelmed and ends up buying too much. Half of it is thrown away at home, anyway. Almost 180 pounds per person per year."

"That's one Marcus a year."

"Excuse me?"

"Marcus weighs around 180 pounds."

"No way, he's lying. He weighs at least—"

"That's not the point."

"You're right. The point is that the supermarkets are too big and sell a bunch of crap that nobody needs but still buys. Probably to comfort themselves. Because the world is so cruel. Hehe. And now we've got these shopping robots. Smart. It'd be good to know whether the stuffed animal department got bigger after they moved into the supermarket."

"Nevertheless, he saved me."

"Who?"

"The handsome guy in the wheelchair."

"Interesting."

"At some point, I wasn't really sure where to go. There was this incredibly soft 'Excuse me, please' coming from all over the place. It sounded more and more threatening. I had to run and dart through the aisles, quickly grabbing what I needed from the shelves."

"Sounds like you got a good workout. But not very disabled-friendly."

"In the end, I dropped everything. But I had to keep running. The robots were on my heels."

"Hehe. At least this will help everyone stay in shape. A lot of movement and very little stuffing your face."

"The sweet man in the wheelchair swiftly grabbed me and pulled me onto his lap. He shifted into a higher gear and brought me safely back outdoors."

"Sounds pretty romantic. I guess you no longer needed the lube for Marcus. But what about the rest of your groceries?"

"We went to the market. Luckily, it was Thursday."

"Cheers, darling, to the future."

# GRAPHIC RECORDING. COOL

---

# VISUAL COMMENTARY BY JOHANNA BENZ

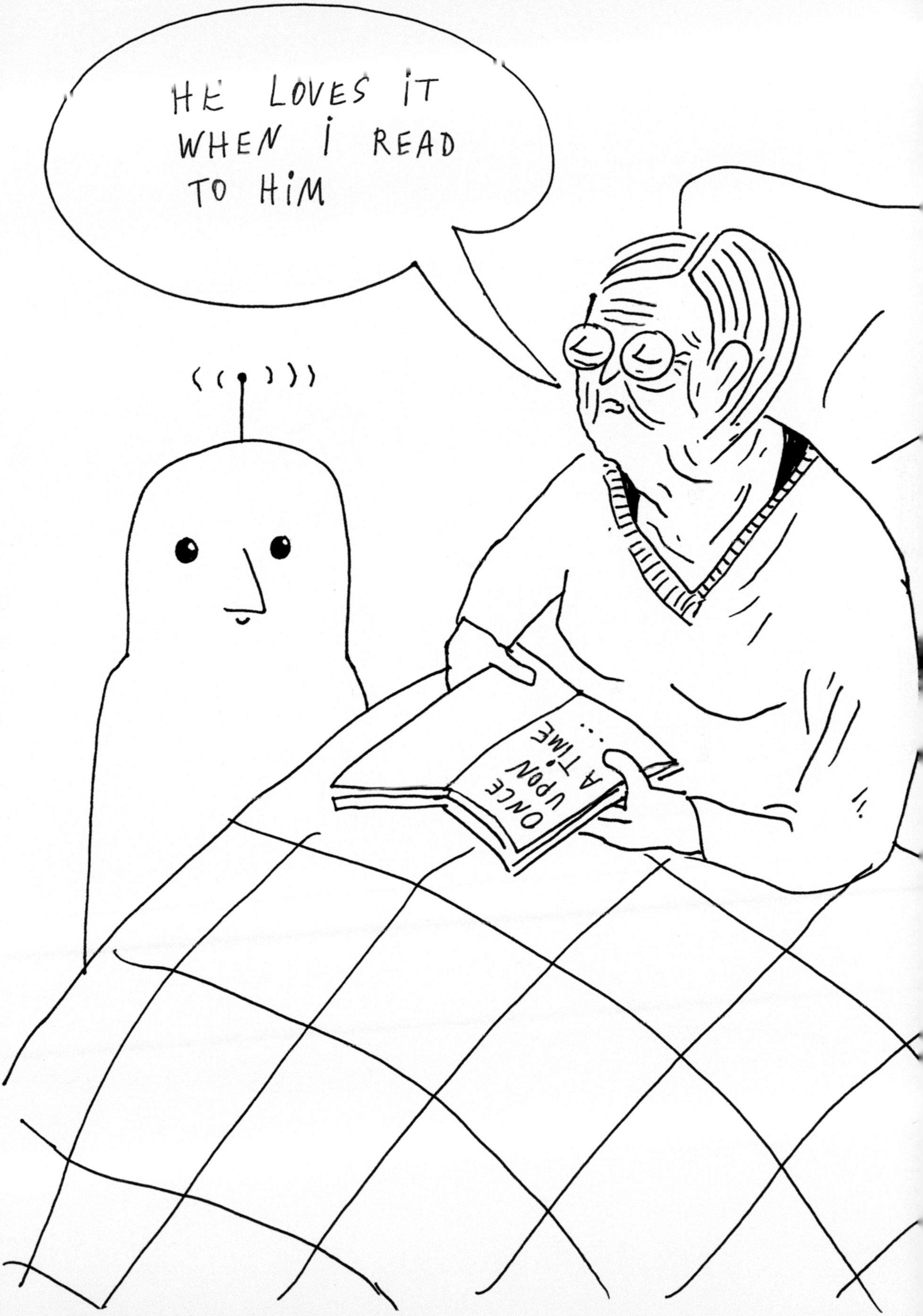
HE LOVES IT WHEN I READ TO HIM
ONCE UPON A TIME...

# DEGREE OF AUTONOMY

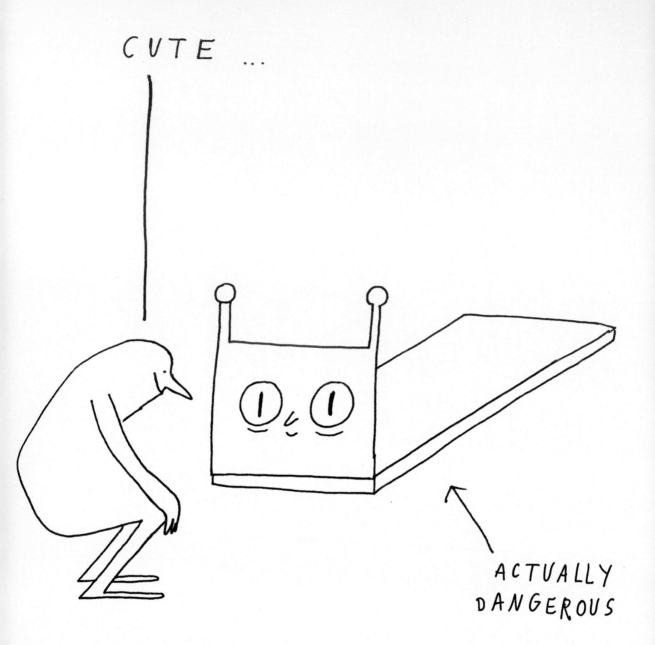

CUTE ...

ACTUALLY
DANGEROUS

ROBOTS SHOULD BE MORE GRUESOME TO INSTILL RESPECT IN HUMANS.

# RECOGNIZING AND ADMINISTERING MASHED POTATOES.

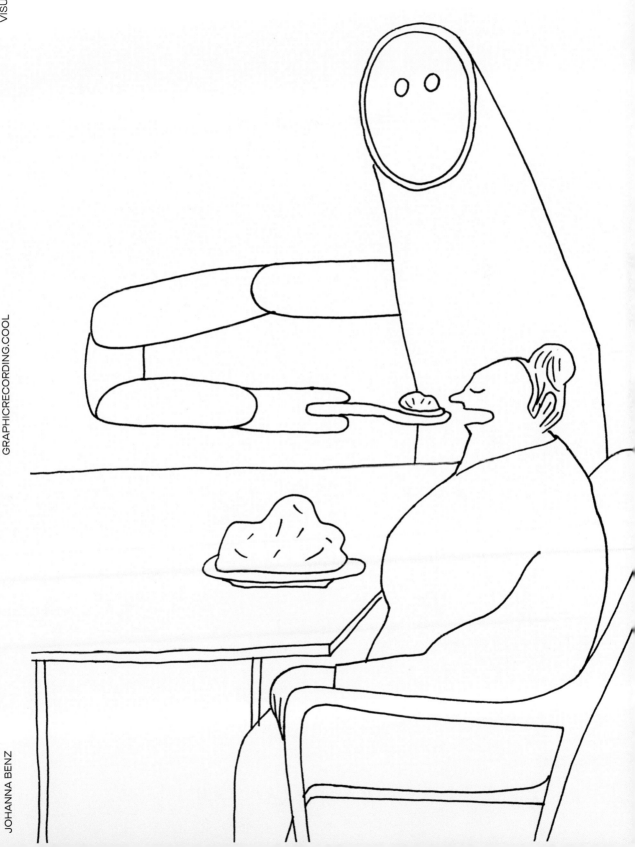

YOU IMPLICITLY HAVE PRIMARY
BIOMETRIC DATA.

AND A PRIVACY PROBLEM.

WE HAVE JUSTIFIABLY CHOSEN A
HUMAN-LIKE APPEARANCE FOR THE
ROBOT.

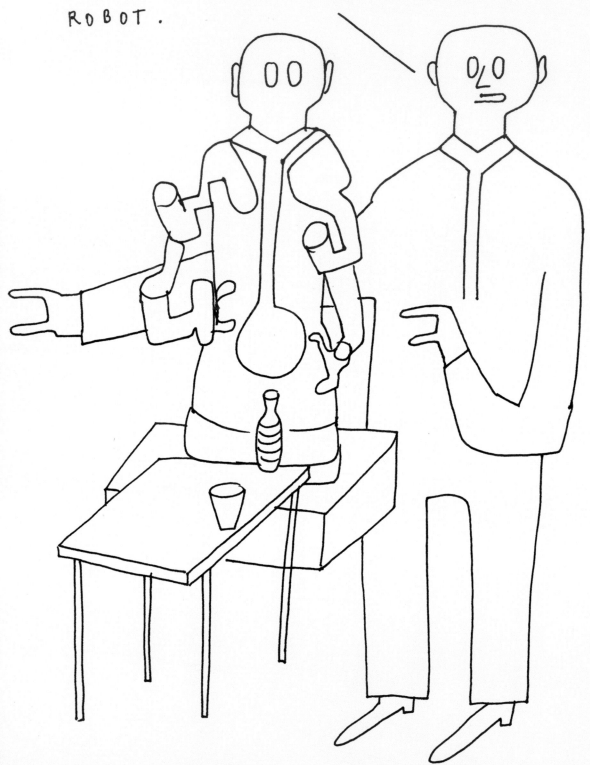

# Part 3
# Designing Together with People—

# Civic Participation and Ethical Implications Concerning Robots

# Citizen Participation in Social Robotics Research

Felix Carros
Johanna Langendorf
Dave Randall
Rainer Wieching
Volker Wulf

If science fiction stories are to be believed, robots will eventually take over the world as artificial intelligence enables a new form of sentient life superior to our own. The truth, however, is much more prosaic. How robots interact with humans now and in the future depends on what they are designed to do, and in turn, the way they are designed will depend on what uses we can put them to. Who gets to decide is of huge importance. Disregarding science fiction, there are other, more realistic scenarios which have great importance for the future of our society. They include such things as the future of the labor force, and with it the financial wellbeing of sections of the population. History is full of examples of designers paying too little attention to, or making errors about, the wants and needs of the public, with consequences which have sometimes been detrimental to society. Developments in weaponry are an obvious example. Another way of putting this is that the design of technology is not simply an engineering problem. It is just as much about the social world and the rights and responsibilities of the people in it. Developing hardware and software which meet the social and other needs of users can be difficult, as users are heterogeneous, their wants and needs vary, and meeting individual needs does not always result in wider social benefits. In the case of social robots, which are the subject of our research, we argue that making sure possible benefits are maximized means getting together with users and other stakeholders to identify what is desirable and what is possible. This is just one example of what is sometimes called "Participatory Design," but which is itself an example of "citizen science." Citizen science in general refers to

IMPULSE⊜ AND TOOLS

CITIZEN PARTICIPATION IN SOCIAL ROBOTICS RESEARCH

FELIX CARROS, JOHANNA LANGENDORF,
DAVE RANDALL, RAINER WIECHING, VOLKER WULF

projects where citizens themselves provide or collect data and possibly analyze the results in collaboration with researchers. There are many benefits. Science itself becomes a more democratic process, because more people have a say in the direction that research should take. Citizens can be a valuable resource, saving time and money in research and bringing new insights with them. Citizen science, then, can be a valuable learning tool for both researchers and citizens themselves.

But what is *citizen science* exactly? In this chapter we give our view, showing how the development of useful social robots can be influenced in beneficial ways by engaging with citizens. In our research we involved citizens in the research process by showing them a robot, demonstrating what it could do and its limitations, and asking them how they might use it and what they would not do with it. The results from these questions influenced the future development of the robot. This gave the development a certain credibility and helped us to balance the different needs people have in relation to social robots. We should emphasize here that social robots are quite different from robots per se. Social robots fulfill a different role than robots have done before. Social robots are developed to interact and work with humans—they do not work in closed factory environments but react to what humans do and say. They are designed for their ability to interact with human beings, to provide responses to users' general behavior, and to answer questions from users in a way that approximates the way humans interact with each other. The fact that it is the quality of interaction that matters is one reason why citizens, or users, have an important role to play.

## WHY CITIZEN SCIENCE?

When social robotics research is conducted within real world settings, potential ethical issues quickly become apparent (Lehmann et al., 2019; Störzinger et al., 2020). These may involve, for example, the responsibilities given to a social robot (e.g., whether a robot should remind people to take their medication) or issues around data privacy (e.g., who gets access to the data produced by a robot's camera).

In this regard, citizen science can be a helpful tool. Citizens are involved in the research process so that researchers can learn from their experiences and reactions. They often have different perspectives on robotics topics, and may see things that the researchers have not thought of simply because they have a pragmatic, real world approach. They will, we can assume, be able to see what robots can and cannot do, and the potential benefits for themselves and others of involving robots in their day-to-day lives. Again, put simply, this acts as a reality check for researchers who might otherwise have quite mistaken assumptions about users' priorities. Showing and discussing use-cases, e.g., using robots to answer older adults' questions in a care home, helps researchers to understand whether interactions of this kind should be encouraged and developed further.

# OUR OWN EXPERIENCES WITH CITIZEN SCIENCE

Our activities with robots in care settings began in 2017, following on from earlier activities related to implementing technology to enhance the wellbeing of residents in care homes. Our research was carried out with the social robot *Pepper*, which has a humanoid appearance, and involved working with older adults and caregivers. In 2018, however, we began to see the need for wider discussions about the implications of social robots, and saw an approach based on citizen science as a way forward. These discussions focused on issues such as whether robots should be part of the German care system, and, if so, how they should be used in that context (e.g., the robots' roles and tasks, and how they could cooperate with care workers). The work with the citizens was part of a → research communication project that was funded by the *German Ministry of Education and Research*, and belonged to a campaign that aimed to foster discussion with citizens about the future of work. As part of the project, we organized over 50 events for citizens, and used these as opportunities to collect views on a range of issues.

Our approach was not to use the citizens as co-researchers, as is often the case, for instance, in climate research, e.g., the European Marine *LitterWatch* project (European Environment Agency, 2021), which allows citizens to report litter around bodies of water. We decided against using this approach because we did not see opportunities where citizens at large could collect data themselves. Rather, our aim was to provide insights into the current technological possibilities for and with citizens, and, based on this, to jointly discuss and develop ideas on how the robot could be used in future care scenarios. This was important for us because social robots may well become a new element in future care, and in this sensitive area there should, we believe, be a debate throughout society about whether and how this new element might be used in future care settings. In this way, it became possible for us to exchange ideas about use scenarios with citizens who are not experts in the field.

**Research Communication:** Every year, the Germany Ministry of Education and Research is funding a Science Year with a specific topic, like future of work, artificial intelligence or bioeconomy. This funding gives researchers the possibility to display their work to a broader public and to discuss it with them.

FELIX CARROS, JOHANNA LANGENDORF,
DAVE RANDALL, RAINER WIECHING, VOLKER WULF

We ultimately tested and combined three different formats for our citizen research:

1. Lectures from experts
2. Discussions with citizens
3. Participatory design workshops

All three approaches were enriching and helped us gain new insights. Our final method, combining the three formats, was organized as follows:

1. Presentation of the robot
2. Presentation of use-cases in care settings
3. Brainstorming on use-cases in care settings; teambuilding
4. Group work: design workshop
5. Presentation and conclusion

At the beginning of each event, we gave a presentation on the current state of our research. This introduced participants to the technology used and the uses of these prototypes in real-world contexts (→ Fig. 1). We then answered initial questions from the participants, such as "How long does the battery last?" or "Can it cook food?". We did this to create a context in which to introduce the opportunities for, and barriers to, this kind of research, and to provide a more realistic view than the one portrayed in the media.

In the case of the design workshops, citizens worked on care robotics concepts in small groups, and the task of the participants was to develop usage concepts based on interactions around a specific task (e.g., the scenario of responding to an emergency alarm, envisaging how such a situation might evolve). The goal was to find desirable applications for the robotic care technology presented, and to describe these by means of text and image flow (see the example in → Fig. 2), and thus to integrate citizens into the design process.

These concepts were then presented to the researchers and other participants and discussed with the group. A diverse range of real-life issues came to the fore, and questions of data protection and ethics were also discussed. The results of the design workshops and the discussions that took place helped us to gain new insights and to understand the different opinions and arguments within this topic. For example, one citizen pointed out that *Pepper*'s design gave the robot an *innocent* look, which could potentially make it easy to use to spy on someone (e.g. care workers).

What was important for us in these events with citizens was that they were based on reciprocity. This means that both sides were expected to take something away with them; citizens should not only be a source of answers to our questions, but should also be given the opportunity to indulge their curiosity. As researchers, we gained knowledge about the application of technology through the exchange, while the citizens gained insights into our research and possible technological solutions for the care sector. Citizens were able to look at the robot, ask questions about it, and receive a detailed presentation of the research results to date. Meanwhile, we were able to gain insights from an important stakeholder group and to further develop our research based on their input.

The results of the citizen participation events helped us to identify a more diverse and general societal picture of care robotics, and to align our research and development with it. A key finding was that robotics should only support the nursing profession rather than replace it (many of the concepts that citizens developed involved three-way interactions between a nurse, a robot, and a person in need of care). This was confirmed in later studies (Carros et al, 2020; Carros 2019). Robots, we found, should not perform automated processes in nursing and elderly care, but should be part of → hybrid teams with nurses or social workers, supporting their work and thus increasing the quality of care.

**Hybrid Teams** are a combination of people and robots working together. Meaning that for the task that should be worked on, both parties are needed to fulfill it (Richert et al., 2016).

# CHALLENGES OF CITIZEN SCIENCE

In all research, particularly research involving citizens, one should be aware of the need to reach out to a variety of groups. As far as possible, all interest groups and a broad spectrum of opinions should be represented. The variety of perspectives one can get from citizens makes it possible to look at ideas from different angles. *Citizen science* thus requires a large network of potential contributing citizens.

In addition, researchers should make their work accessible for citizens, in order to create awareness of the research field. Scientific publications often have a complicated structure and use a language all of their own. This style of writing can make understanding difficult for readers who are not familiar with the subject at hand. Scientific research is often prepared only for the scientific community and not for the public, although in most cases it is financed by the public. This has important implications, since it is only reasonable that citizens who both pay for innovation through their taxes and social security payments, and who may directly experience the effects of robotics, should have some say in how robotics develops.

Robotics research has only recently paid attention to the *interactive* or *social potential* of robots. This is changing, and we can anticipate that robots will become increasingly present in everyday life, which is why citizens' understanding of social robots, and their attitudes towards them, matter. Thus, a broad network of citizens interested in science would bring societal benefits for all the reasons

FELIX CARROS, JOHANNA LANGENDORF, DAVE RANDALL, RAINER WIECHING, VOLKER WULF

mentioned above. Making sure that the views collected are representative, however, is not easy. Even so, although this means more effort for researchers, it is important for the future of the research field in the long term, as it improves public understanding of science. One challenge is the fact that citizen involvement in science is often not undertaken systemically within research projects. Public relations often play a rather subordinate role, despite efforts by the various funding agencies, and tend to relate to the publication of results rather than to the participation of citizens in the research process. This problem has already been recognized but not yet solved, despite specific research communication projects such as the "Science Year" in Germany, and Europe-wide platforms created as a basis for citizen science projects (European Commission, 2021).

Events involving citizens, we believe, can and should be mutually beneficial. Researchers gain from the insights that a wide range of inputs can bring, and citizens gain a better understanding of what is possible and what might be desirable. Presenting the state of the art in research and development in robotics to the citizen can be challenging, but can nevertheless result in something which benefits everyone. This *giving back*, if done in the right way, repays the effort of researchers to engage citizens by providing insights which are closer to real life scenarios, thus improving our picture of what the social factors involved in the use of robots might be.

In our experience, it is important to be aware that citizen science projects deliver good and practical results when citizens are involved in the entire research process, not just specific parts of the research. This argument is also fundamental to a concept that has recently become associated with citizen science in Europe, the concept of "Responsible Research and Innovation" (RRI). The core idea here is that societal stakeholders should be involved throughout the research process to incorporate society's values, needs, and expectations into results (European Comission, 2013). Ultimately, public trust in science requires us to move away from the idea of the *absent minded professor* working away in a lab, wearing a dirty lab coat as a result of too many experiments. A clear way in which we can build this trust, and enhance mutual knowledge, is through citizen science, in whatever form this takes.

# CITIZEN SCIENCE—FURTHER METHODS AND RESOURCES

There are various ways to conduct citizen science. In the last paragraph we explained our approach, and here we want to present alternative approaches and resources that could be helpful. For example, the website *Citizens Create Knowledge (Bürger Schaffen Wissen,* www.buergerschaffenwissen.de) was launched to develop such projects. It not only provides comprehensive information about relevant topics, but also helps working groups and networks start citizen science projects which can be published on the platform. It also enables researchers to participate in other projects, allowing them to gain insights into other work going on in their field which may inform their own work. The website also allows researchers to access an existing network of German-speaking citizens.

While *Citizens Create Knowledge* is a German platform, the *European Citizen Science Association (ECSA)* offers a similar service for EU-wide projects (→ Fig. 3). *ECSA* provides several options for participating in research projects, for example, as a research partner, as a communication partner or by hosting events and workshops. Depending on the required expertise and experience, *ECSA* can contribute to projects both centrally through *ECSA* headquarters and through its network of members (European Citizen Science Association, 2021).

Another method for approaching citizen science is presented by Kirstin Oswald in her article *Citizen Science on Instagram*. She focuses on the social media platform *Instagram*, particularly the use of hashtags, to evaluate public attitudes to citizen science (Oswald, 2021). In general, social media can be an important and productive research tool for citizen science. Ambrose-Oji et al. (2014) demonstrate this using several case studies in environmental research in their report *Citizen Science: Social Media as a Supporting Tool*. In doing so, they not only consider social media as a research tool, but also look carefully at the data quality and collection methods in research conducted through social media.

Citizen science requires networking and exchange between science and society (Bürger Schaffen Wissen, 2016). Both social media and websites such as those presented above can be tools for this. They can also help citizen science become more widely recognized in society and the science world (Bürger Schaffen Wissen, 2016). For successful collaborative research, it is also necessary to integrate citizen science into scientific processes and into educational concepts, as well as integrating its results into decision-making processes (Bürger Schaffen Wissen, 2016).

Ambrose-Oji, B., van der Jagt, A.P.N., & O'Neil, S. (2014). *Citizen science: Social media as a sup-porting tool*. Forest Research.

Bürger Schaffen Wissen (2016). *Grünbuch Citizen Science Strategie 2020 für Deutschland*. https://www.buergerschaffenwissen.de/sites/default/files/assets/dokumente/gewiss-gruenbuch_citizen_science_strategie.pdf

Carros, F. (2019). Roboter in der Pflege, ein Schreckgespenst? *Mensch und Computer 2019-Workshopband*.

CITIZEN PARTICIPATION IN SOCIAL ROBOTICS RESEARCH

**Fig. 1**  Explaining the functions of the robot. © University of Siegen, Information Systems and New Media

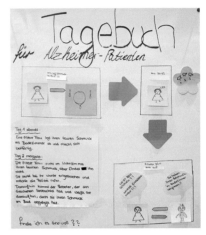

**Fig. 2**  Example of an application area developed by citizen in a design workshop. © University of Siegen, Information Systems and New Media

FELIX CARROS, JOHANNA LANGENDORF, DAVE RANDALL, RAINER WIECHING, VOLKER WULF

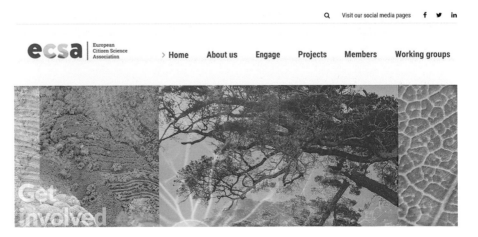

**Fig.3**  The ECSA website

Carros, F., Meurer, J., Löffler, D., Unbehaun, D., Matthies, S., Koch, I., Wieching, R., Randall, D., & Hassenzahl, M. (2020). Exploring human-robot interaction with the elderly: Results from a ten-week case study in a care home. *Proceedings of the SIGCHI Conference on Human Factors in Computing Systems CHI 2020. Association for Computing Machinery, 1–12.* https://doi.org/10.1145/3313831.3376402

European Citizen Science Association (2021). *European Citizen Science Association*. https://ecsa.citizen-science.net/

European Commission, Directorate-General for Research and Innovation, (2013). *Options for strengthening responsible research and innovation: report of the Expert Group on the State of Art in Europe on Responsible Research and Innovation*, Publications Office. https://data.europa.eu/doi/10.2777/46253

European Commission (2021). Citizen science. https://ec.europa.eu/digital-single-market/en/citizen-science

European Environment Agency (2021). *Marine LitterWatch*. https://www.eea.europa.eu/themes/water/europes-seas-and-coasts/assessments/marine-litterwatch

Lehmann, J., Carros, F., Unbehaun, D., Wieching, R., & Lüssem, J. (2019). Einsatzfelder der sozia-len Robotik in der Pflege. In Nicolas Krämer and Christian Stoffers, Eds. Digitale Transfor-mation im Krankenhaus. Thesen, Potenziale, Anwendungen. Mediengruppe Oberfranken, Kulmbach, 88–113.

Oswald, K. (2021). *Citizen Science auf Instagram*. https://bkw.hypotheses.org/1969#more-1969

Richert, A., Shehadeh, M., Müller, S., Schröder, S., & Jeschke, S. (2016). Robotic workmates: Hybrid human-robot-teams in the industry 4.0. In *International Conference on e-Learning* (p. 127). Academic Conferences International Limited.

Störzinger, T., Carros, F., Wierling, A., Misselhorn, C., & Wieching, R. (2020). Categorizing social robots with respect to dimensions relevant to ethical, social and legal implications. i-com, 19(1), 47–57. https://doi.org/10.1515/icom-2020-0005

IMPULSES AND TOOLS

CITIZEN PARTICIPATION IN SOCIAL ROBOTICS RESEARCH

FELIX CARROS, JOHANNA LANGENDORF,
DAVE RANDALL, RAINER WIECHING, VOLKER WULF

# KoBo34

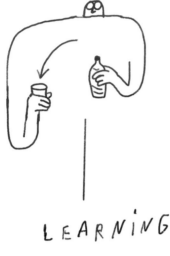

LEARNING

## AIM OF OUR RESEARCH PROJECT

We aim to support older people living in nursing homes to keep their independence and self-determination by allowing a robot to help them with specific tasks in their daily lives. We identified these tasks through an extensive needs analysis with the residents, their relatives, nursing home staff, the management of the care facility and senior citizens living in their own homes.

## CONTEXT, ROLE AND TASK OF OUR ROBOT

The robot is a servant/assistant that supports older people in their daily lives through a safe and intuitive haptic interaction, especially for two-arm tasks, such as serving food and clearing tables, offering drinks and snacks, and helping with shopping.

## MOST ASTONISHING FINDING LEARNED ABOUT ROBOTS DURING OUR WORK

Residents and nursing home staff had several of their own ideas about how technology could be helpful for them. We saw that it was important to respect their fears and concerns in order to achieve a good result. Our own internship within the care facility gave us a better understanding of the needs and spatial conditions that we take into account.

## WHO WE ARE

TH Rosenheim, Faculty of Applied Health and Social Sciences and Centre for Research, Development and Technology Transfer; TU München, Institute of Robotics and Machine Intelligence; TU Darmstadt, Intelligent Autonomous Systems; TU Darmstadt, Psychology of Information Processing; TU Darmstadt, Center for Cognitive Science; Franka Emika GmbH; Rummelsberger Diakonie e.V.

# Learning from Each Other—How Roboticists Learn from Users and How Users Teach Their Robots

## Interview with

Eva Theresa Jahn (E.J.)
Dorothea Koert (D.K.)
Susanne Trick (S.T.)
Martin Müller (M.M.)
Caren Horstmannshoff (C.H.)
Christoph Jähne (C.J.)

## By

Felix Carros
Adrian Preussner

WITH KOBO34, YOU HAVE DEVELOPED A HUMANOID SERVICE ROBOT THAT IS INTENDED TO SUPPORT PEOPLE IN NEED OF NURSING CARE IN THEIR EVERYDAY LIVES WITHIN A NURSING HOME. HOWEVER, YOUR ROOTS ARE PARTLY FROM DIFFERENT DISCIPLINES, SO YOUR KNOWLEDGE OF NURSING IS RELATIVELY LIMITED. HOW DID YOU ARRIVE AT YOUR FINDINGS?

E.J.     On the one hand, we conducted an extensive needs analysis at the beginning of the project with our partner, the Lenzheim nursing home's residents and nursing home staff. In addition, they allowed us to do an internship with six people from the research project for a duration of ten days. The experience was very valuable for all of us, since most of the researchers had no prior experience in the field of nursing.

**D.K.** It helped a lot that we were able to visit Lenzheim at the beginning of the project so we could get a much better idea of what is actually done on site in everyday life and what might be needed. The composition of our research group was also a clear added value here. The exchange with the *Rosenheim Technical University of Applied Sciences* (nursing science research) always helped us a lot from the *technical side,* I think, to ensure that our ideas did not stray too far from what makes sense in reality.

**WHAT WAS THE MOTIVATION FOR THE INTERNSHIP? WHAT DID YOU HOPE TO GAIN FROM IT?**

**M.M.** Understanding the field is important for user-centered and needs-oriented research and development. We planned the internship to ensure that scientists from other disciplines are also able to understand the context of the nursing home, the processes and its stakeholders. Additionally, we aimed to build a common understanding and a common language within the research team. In addition, the interns wrote short protocols, which we included in the "needs" assessment.

**S.T.** In our research, we usually abstract and simplify the application examples in order to make them technically feasible. Contrasting this, the internship gave us a realistic insight to what is really needed and what will be implementable on site. In addition, it was a perfect opportunity to have a discussion with the people concerned and to communicate, for example, that we do not want to replace human workers, but to support them in the best possible way.

**D.K.** I think that one can only develop something meaningful if they have a clear understanding of the target application and environment. The internship also helped me to imagine not only something abstract, but the people in the Lenzheim nursing home as a target group.

**HOW DID YOU ORGANIZE YOURSELF AS A RESEARCH PROJECT FOR THE INTERNSHIP IN THE CARE HOME?**

**C.J.** In the research group and together with the nursing home, we drew up a schedule in which one person was assigned to the early or late shift for two weeks at a time. We were not allowed to help much with most of the activities. For many things, training as a nurse would have been necessary for insurance reasons, so our role was limited to observing and accompanying the work.

**DID YOU ALSO LEND A HAND AND HELP?**

**E.J.** We completely took on the role of interns. Our early shift began at 6 a.m., and we entered the nursing home in our ward clothes. We were assigned to a trained nurse and helped to complete their daily program. Wherever there were tasks that did not require specific training, we helped. For example, we handed out small objects, assisted with meals, or held conversations.

**WHAT DO YOU REMEMBER MOST ABOUT IT?**

**E.J.** After half a day you feel exhausted, but also satisfied because you have helped so many people.

**S.T.** I fully agree with Eva!

**HOW DO YOU THINK THE FINDINGS WOULD HAVE DIFFERED IF YOU HAD NOT DONE AN INTERNSHIP?**

**C.H.** The basis of our cooperation would have been different. There are many reports about the work requirements in nursing homes, like time pressure. But when you experience it by yourself and conversations can take place, it becomes easier to have a better understanding of the field. It is also important for future tasks that may go beyond this project. No one will forget the impressions and conversations.

**M.M.** I don't know to what extent the findings would have been different. However, I assume that the understanding of the problems became much clearer to all those who were involved, and that user-centeredness was more successful. What was achieved, in any case, was a feeling of unity within the very heterogeneous team and a good starting point for a trusting cooperation.

**S.T.** Apart from technical specifications and the definition of target scenarios, I think it is always particularly important in research to have a vision of what the research will be used for, especially because it doesn't directly result in a product. For this vision, the internship was very helpful for me.

**NOW WE'VE TALKED A LOT ABOUT HOW YOU LEARN AS RESEARCHERS AND GET TO KNOW THE SETTING, BUT HOW DOES THE ROBOT ACTUALLY LEARN? DOES IT ALSO HAVE TO DO AN INTERNSHIP?**

S.T.    That would indeed be a dream goal to have the robot walk along with a nurse, and by doing so, learn the skills to support them. This is not yet feasible, but we are working towards this goal, for example, by enabling a human to teach a task to a robot by demonstration. Another way is to give feedback on the robot's actions in order to make the system learn from its mistakes.

C.H.    Even the robot would need to do several internships and to always keep learning something new. Things that you only think about in the lab don't necessarily work with the participants.

D.K.    The idea is that the robot should eventually be able to learn from people who have no prior experience with robots and also limited experience with technology. The long-term goal would be for it to be able to learn from natural demonstrations, that is by being shown things. In the *KoBo34* project, we are taking the first step by using a graphical user interface (for example, a tablet) to compose new tasks from known components and to show it new movements by *taking it by the hand* and guiding it.

EVA THERESA JAHN, DOROTHEA KOERT, SUSANNE TRICK,
MARTIN MÜLLER, CAREN HORSTMANNSHOFF, CHRISTOPH JÄHNE

# ROBOTTTT

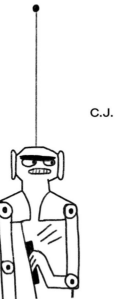

**HOW DO YOU GET THE NURSING STAFF TO TEACH THE ROBOT SOMETHING, AND TO WHAT EXTENT IS THE ROLE OF THE NURSING STAFF CHANGING BECAUSE OF IT?**

**S.T.** We are trying to make the interaction between the robot and the nurse as intuitive as possible. Special technical skills (such as programming) are therefore not necessary. Nevertheless, working with robots will change nurses' everyday work, for the better, I hope. In any case, it should be avoided that society gets the impression that humans can be replaced by robots for financial reasons, which is a possible risk of using robots in nursing. However, especially after our internship in a nursing home, it is clear to me that a robot cannot replace the work done there, it should only support it in the best possible way. So, with my research I want to contribute to a sensible use of robots in nursing.

**E.J.** I think that this will definitely make work more colorful and hopefully a little more relaxed for the nursing staff if some of the simpler and more routine service tasks can be transferred to the robot.

**WHY DO YOU WANT SOMEONE ELSE TO TEACH THE ROBOT SOMETHING? WHY DID YOU DECIDE TO TAKE THIS APPROACH?**

**C.J.** Every application is different. The versatility of the environments in which the robot should be able to provide meaningful assistance makes it necessary to be able to adapt. The use of the robot by nursing staff in particular can also promote acceptance and hands over both control and responsibility to an important extent to the specialist staff. The robot will not become independent in the foreseeable future, hence the image of a tool used by humans.

**BECAUSE SOMEONE COULD TEACH THE ROBOT TO DO SOMETHING ILLICIT OR UNWANTED, LIKE <u>STEAL</u> OR BRING THE <u>INCORRECT MEDICATION</u>, ISN'T IT DANGEROUS TO DEPEND ON IT? OR COULD THE ROBOT LEARN ON ITS OWN?**

**E.J.** I believe that there will be a clear separation as to how far nursing staff or other responsible persons can train the robot. That is the responsibility of the developers, but of course it is also in their interest. No one would place their trust in a robot that can be completely manipulated.

**HOW LONG DO YOU THINK IT WILL TAKE UNTIL** ROBOTS **HAVE LEARNED ENOUGH TO BECOME COMMONPLACE IN NURSING HOMES?**

S.T.  I would return the question, when will humans have learned enough? In this respect, I would say that the more the robot learns, the better, because then it will be able to contribute more and more to supporting nurses, or humans in general. I think it will still take a few years before robots are used across the board and are helpful enough to become commonplace in nursing homes. Much of what works in this area is currently still very much tied to stable laboratory conditions and quickly fails in real-life scenarios, but we are working on that!

C.J.  I estimate that it will be another 15 years before service robots play a role in German nursing homes. The impression conveyed by fascinating videos about the state-of-the-art in robotics is deceptive. The legal framework has yet to emerge. For this, decision makers must first understand such technology to a sufficient extent to be able to create meaningful laws for it and the use of such robots will not be implemented without health insurance companies' approval. Not only that, but they must also first be shown (and rightly so) that a nursing robot can bring real added value. This added value has yet to be developed, so it remains exciting.

EVA THERESA JAHN, DOROTHEA KOERT, SUSANNE TRICK,
MARTIN MÜLLER, CAREN HORSTMANNSHOFF, CHRISTOPH JÄHNE

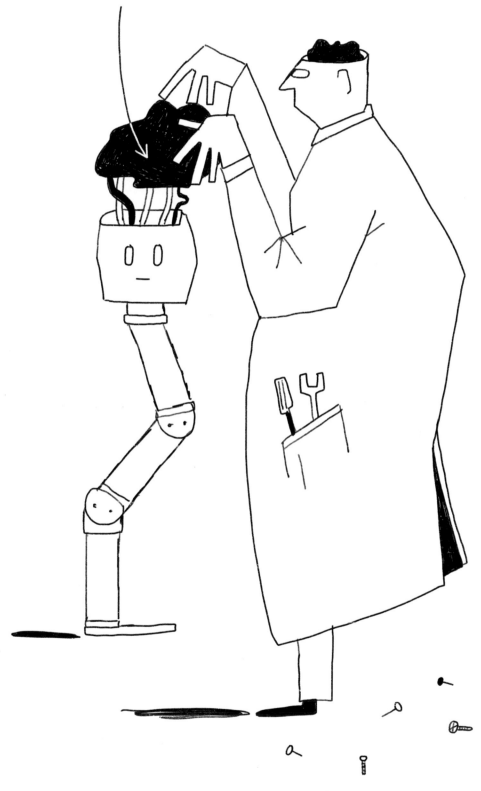

LIFELONG LEARNING

Inspired by the interview, the illustrator Johanna Benz visually
commented on chances, risks and scenarios of robotics from her
own perspective. © Johanna Benz

## Edi Haug

lives near Berlin and catches most of his Pokémon in Prague,
London and the south of Berlin, with the help of his human
friend, Laura, or his family. Due to a hereditary disease, he has
lived deaf and blind since he was 12 years old. He is currently
studying shorthand braille and SMS writing, and he wants to
become a masseur.

## Laura M. Schwengber

is a Sign Language Interpreter, has an MA in barrier-free
communication and is writing her PhD at the *Institute for
Easy and Plain Language* at *University of Hildesheim*. Edi has
been her best friend for 20 years now. She lives in Berlin
and teaches people how to write barrier-free social media
postings with her team #BarriererfeiPosten.

# My Friend *Simsala*, the Robot

Edi Haug
Laura M. Schwengber

Hi, my name is Edi and I live deaf and blind. I am dictating this text to my best friend, Laura. I could write this text myself by typing it into my *Braille Wave*, which is connected to my computer or my *iPhone*. A *Braille Wave* is a kind of electronic display and typewriter for braille. It shows the text written on a screen in braille so I can read it with my hands. Furthermore, I could use *Siri* to write my spoken words down. But as I have been hard-of-hearing since I was 10 years old, my pronunciation is a little hard to understand and this text would not be understandable at all, rather just a little weird. I would produce a barrier—or, my friend, *Siri*, would cause it. So, long story short, I asked Laura to assist me. Laura is, unlike *Braille Wave* and *Siri*, a human friend. Laura is able to read between the lines, she is full of empathy, and often I do not have to tell her what I want; she already knows. I wish a technical tool could be a bit like her, but neither *Braille Wave* nor *Siri* is.

Using the *Braille Wave* on a computer or an *iPhone*, you would be able to communicate with me too. We would chat, or you would have to learn the Lorm alphabet and sign every letter into the palm of my hand and I'd respond to you. No technics, just you learning the 26 letters of the Lorm alphabet, because it is super easy to learn!

But to be honest, would you learn the Lorm alphabet just for me to be able to order a vegetarian burger at your restaurant only once a month? Would you learn to use it fast enough to understand me for when I fall off something and break a leg, or when I have a silly accident because something in our surroundings isn't accessible? Would you want to touch my sticky hands in the cinema as I eat yummy popcorn, and you are translating the movie for me? Yes, I go to cinemas and yes, sweet popcorn is the better one.

Communication is key when it comes to me living deaf and blind, but it doesn't end there. It is me trying to wake up in the morning on time, it is me trying to find my left shoe, and it is me traveling the world with my *Pokémon Go* game in one and—well, what? Who? in the other hand. Until now, you'd find Laura occupying this very special place in my other hand. She is my friend but, at the same time, she is my assistant, my interpreter, and my guide all in one. And from time to time, this is way too much for her to handle. We dream of a life where Laura can be just my friend, period, and where this very special place in my hand can still be my means of communication, my means of exploring the world, my means of saying hi to you and hi to *Simsala*.

*Simsala* would be my robotic friend, guiding me to the next burger store, ordering the vegetarian one, helping me find my shoes and telling me about the latest films at the cinemas so I can go watch them with Laura and she does not have to be a kind of tool anymore. And if you wanted to meet me, you'd have the choice between my sticky popcorn hands or *Simsala* to communicate with me. Further, sometimes I just want to do things on my own. Take a shower, wash my underwear, have an extra unhealthy second serving of cheese and stuff on my third burger within a week, without anyone judging.

However, people decided to build a world where there are millions of barriers for me when it comes to living a self-determined life, so I like the idea of rebuilding it. Whosoever is in charge, tear down these walls! As you probably know, I am not the only deaf and blind person on this planet and, holy moly, there are so many of us living a limited version of our lives because of the world we are living in, not because of my, or others', individual handicaps. So, who am I to demand barrier-free surroundings while the society around us shows this extraordinary level of ignorance? It is obvious: a lot of people do not care about me (or us) living a self-determined life. If the majority of people truly cared, my *Pokémon Go* game would show more than one single dot on my *Braille Wave*, but apparently it wasn't programmed for blind people anyhow. This lonely dot is the big company's, and society's, way of saying: "hi, we don't care," or at least that they don't care enough to invest in programming an accessible game for everyone—so, thanks for saying hi.

Pretty devastating, isn't it? But—by all means there are people who do care. There are companies who do invest in accessibility. I have a *Braille Wave*, I get paid a budget for human assistance every month, I have a vibrating watch waking me up in the morning. There are those people—thanks for saying hi, too.

Accessibility is key. For me, a still speaking, but not hearing, blind man who can't stand for longer than 15 minutes in a row, accessibility can be a chair, braille or stuff that vibrates. For a person with the exact same handicaps, it can be something very different. I don't need a kitchen scale with braille—I don't cook. Instead, I need accessible food delivery apps. Long story short: we will not end up in a fully inclusive world. There will always be barriers, for me and for others. At least, in a society like ours, I will not live long enough to experience a world where I can travel on my own with all information in braille, experience every single pathway equipped with a guidance system for the blind, or be able to order a latte with extra flavors at any given coffee shop in Lorm. But what if I am not meant to wait for this never-to-happen perfect accessible world? I mean, I will not be waiting for it. I will call Laura and we will go change the world by us traveling around, talking to people, talking to the *Pokémon Go* guys (call me back!), teaching you Lorm, and answering all your questions concerning accessibility. What do I need from you? Ask me, mail me, listen to me.

Together, we would work at creating *Simsala*, and I would like you to meet her, she would be my robot. She already has a name and, in my mind, she is a perfectly functioning, magically coffee brewing, shoe-finding, underwear-washing, never-out-of battery robot-friend. The world isn't accessible—so what? Build me something that magically bridges that gap. I have no clue at all about robotics, about how hard developing a *Simsala* would be, but I am sure you can do that! The frustrating thing for both you and me will be that no matter how super-duper perfect *Simsala* might be, she will never be enough, and never be more than a tool. Neither you nor I will ever build a perfectly inclusive and accessible world for myself and all of us. Nonetheless, feeling accepted, appreciated, and feeling included is a thing—it is huge. I am looking forward to meeting all your robot-friends and their developers, who show me that they do care. A lot. Never think of your robots as pure technical products or machines—you are working with friends. Be brave, include me!

# MIRobO

### AIM OF OUR RESEARCH PROJECT

We aim to design a safe and intuitive robot that can hand over objects to a broad range of people in an integrative way by using different modalities, like speech and haptic feedback.

### CONTEXT, ROLE AND TASK OF OUR ROBOT

The mobile robot of *MIRobO* is a robotic arm handing over potentially dangerous objects, like a filled cup or a knife. Its interaction partners are sighted people who can see the robot and objects, as well as people who are visually impaired or blind. As an example for context, we chose a hospital scenario.

### MOST ASTONISHING FINDING LEARNED ABOUT ROBOTS DURING OUR WORK

Users perceive the task of handing something over as trivial; therefore, they state performance expectations that are very high. They call for broader and more useful functionalities.

### WHO WE ARE

Chemnitz University of Technology, Professorship Ergonomics and Innovation Management and Private Law and Intellectual Property Rights; HFC Human-Factors-Consult GmbH; YOUSE GmbH; Fraunhofer Institute for Machine Tools and Forming Technology; FusionSystems GmbH; Sikom Software GmbH.

# Move Away from the Stereotypical User in the Picture-Perfect Scenario—A Plea for Early and Broad User Integration

*Interview with*

Kai-Uwe Kaden (K.K.)
Daniel Wimpff (D.W.)
Dorothea Langer (D.L.)

*By*

Stephanie Häusler-Weiss
Kilian Röhm
Tobias Störzinger

**WHICH SCENARIO IDEAS DID YOU START THE PROJECT WITH?**

D.L.　To come up with a fitting interaction concept for the robot, our goal was to develop a scenario based on user surveys at the beginning of the project. According to our prior knowledge, the interaction should take place multimodally, that is, via several sensory channels. This is technically safer, and people do not communicate via one single sensory channel. Accordingly, our idea was that the interaction with the user should happen via speech and gesture control.

**HOW DID YOU SPECIFICALLY GO ABOUT INVOLVING USERS?**

D.L.   First, we thought about the interaction conceptually, and at the same time tried to get in touch with users who are visually impaired. This was not easy, but once we had a foot in the door, they told each other about our project, which was a great help to get in touch with more users. As a result, we were able to conduct a focus group with people who have limited and full vision, in which we asked mainly everyday questions. For example, we asked how the interaction of handing over objects works between two people, what is difficult or easy for them and so on. Blind participants also reported where they need support and what this support would ideally look like. Against this backdrop, we then started a series of in-depth interviews in which we called upon other blind people and ultimately derived ideas from all this information.

**BUT PEOPLE WITHOUT SPECIAL NEEDS PARTICIPATED IN YOUR FOCUS GROUPS TOO. WHY?**

D.L.   Our approach was a working interaction concept for everyone and not a special solution for people with disabilities. Later user studies showed that fully sighted people can also benefit from our findings.

**WHAT EXPERIENCES DID YOU HAVE PREPARING AND CONDUCTING THE FOCUS GROUPS?**

D.L.   At first, we contemplated whether the standard procedure of conducting a focus group was applicable at all. For the most part, we stuck with it and were happy to have supporters with us who, for example, wrote down the blind participants' ideas. We were flexible in our choice of methods and spontaneously saw what was feasible and what was not, but we always had a backup ready. One shouldn't be afraid to work with a special needs user group. As an ice-breaker, instead of visual illustrations, we passed around little filled bags as a tactile riddle. It was really amazing how quickly the blind participants guessed the items! In retrospect, the extra effort was a lot less than expected and the valuable feedback made it worth it.

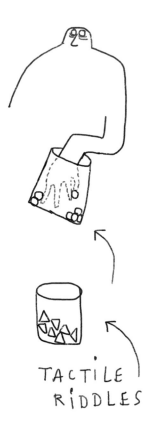

TACTILE RIDDLES

## WHAT WERE INSIGHTS YOU GAINED FROM USER ENGAGEMENT?

**K.K.**  It turned out that handing over objects in the air is fundamentally challenging because of the lack of orientation. The finding, which was also confirmed in the interviews, was that placing the object on a surface whenever possible was preferred. So, handing over and picking up something should be separate actions, but there are also situations when handing something over in the air is inevitable. For example, participants reported a situation at the Christmas market where handing over a hot bratwurst or mulled wine sometimes causes problems. In these cases, additional signals for orientation are helpful.

**D.L.**  Another topic was that blind users generally rejected our idea of using gestures when being handed something by a robot. Although they were able to show purposeful gestures in experiments, they said they would not perform any gestures without knowing exactly where the robot with the potentially dangerous object was. Because of this, we massively reduced gesture communication. Together with blind people we have developed a few possible gestures, but their evaluation is still pending. Our blind participants would prefer a remote control or an app for alternative control besides speech, but definitely <u>not gestures</u>.

**D.W.**  A consistent need among blind people was that only one person at a time should be active during an interaction. If handing something over that involves a risk of injury, they definitely want to be in control. Beyond that, they would expect the sighted partner to take the active role. However, they were unsure whether they would feel the same way with a robot. This still needs to be evaluated.

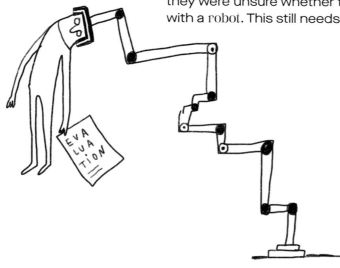

**HOW DID THE RESULTS OF THE USER SURVEYS AFFECT THE SCENARIO? WHAT KIND OF SCENARIO WAS DEVELOPED AFTERWARDS?**

**D.L.** A very artificial scenario was created since there was clear feedback from the user group that help and support in everyday situations is most likely needed in unfamiliar environments. Focusing on communication between the participant and the robot, we chose the following scenario: a bedridden, blind patient in a hospital. The person is both restricted in actions and also situated in an unfamiliar environment. In fact, the person would have to call the nurse for any help, for example, to have something brought to their bedside. Importantly, the user surveys revealed that blind people do not want objects to be handed over in the air. Therefore, the scenario was refined so that the robot can also put down the objects in a fixed place, like on the table next to the patient's bed, and the patient would then be able to pick it up from there.

**IN YOUR PROJECT, YOU ALSO TOOK A DETAILED LOOK AT DIFFERENT BEHAVIORS THAT INVOLVED GRABBING. HOW DOES THE BEHAVIOR OF BLIND AND SIGHTED PEOPLE DIFFER?**

**D.L.** We compared the grabbing behavior (poses, speeds, etc.) of 60 participants: 20 blind, 20 sighted and 20 with blindfolds. Here, it was particularly interesting to see that the blind participants behaved differently from the sighted participants wearing blindfolds. This means that the behavior of user groups with special needs cannot simply be simulated but must be directly included. Later, it became apparent that the solution which is usable for blind people can also be used by the fully sighted without any problems.

**WHAT WOULD YOU WANT TO SHARE WITH PROJECTS THAT ARE JUST STARTING AND MAY BE DEALING WITH A USER GROUP WITH SPECIAL NEEDS?**

**D.L.** It is important to get in contact at an early stage and to keep contact regularly in order to attain incredibly valuable feedback. Basically, this should already be happening while creating the usage scenario and describing the problem. In addition, by involving user groups with special needs in the further development of the concept, one can always get feedback on whether a solution is suitable or not.

A distinction can be made between the two phases of problem definition and iterative testing: while it is good to have many perspectives during problem definition—a size of about 20 participants is suitable here—it makes sense for development to use small steps and proceed iteratively. In the early phases, large-scale testing is likely to be unnecessary. For example, if an evaluation with 50 participants is carried out too early, dead ends will be encountered because the system is just too buggy at this stage. It is better to test iteratively with two or three people over and over again.

**WHAT ELSE IS IMPORTANT FOR THE DEVELOPMENT PROCESS?**

**D.L.** The development of technology can quickly become exclusionary. Our plea here is to be inclusive and to move away from the stereotypical standard user in the *picture-perfect scenario*. Much more work needs to be done to make basic robot skills robust for diverse user groups.

Inspired by the interview, the illustrator Johanna Benz visually commented on chances, risks and scenarios of robotics from her own perspective. © Johanna Benz

# Is it Good?—A Philosophical Approach Towards Ethics-Centered Design

Catrin Misselhorn
Manuel Scheidegger
Tobias Störzinger

Robots, autonomous driving, artificial intelligence and big data are all examples of technologies that increasingly call for ethical reflection regarding their impact on our society.

Ethical considerations have thus been an essential part of the development of technology for quite some time now. Scientists, engineers and designers have learned that technology is never neutral, but embodies certain values as well as creates new horizons for actions that can lead to negative or positive outcomes (see, e.g., Cummins, 2006; Grunwald, 1999, 2013; Misselhorn, 2018). Scientists react to the ongoing and rapid progress of certain technologies (like robotics) with new research disciplines (for example, the emergence of →machine ethics or "emotional AI" as new disciplines, see Misselhorn, 2018; 2021).

Everyone seems to agree that ethical considerations play an important role when it comes to designing and implementing technologies. Large corporations are beginning to spend large amounts of money in order to reflect upon potential ethical problems and risks (Mickle, 2021). While this is positive news for everyone who cares about the future, we argue that the true potential of incorporating ethics into technology has not yet been met. There are two reasons

**Machine Ethics** addresses questions such as whether machines can act morally at all, whether they may act in morally sensitive situations, or whether they can be patients of moral actions i.e., beings that have to be taken morally into consideration without necessarily being moral agents. Machine ethics is a new discipline at the interface of philosophy, computer science and robotics.

for this. First, the most widely used ethical concepts in technology focus solely on negative aspects and harm prevention. They presuppose that the positive value of technology is solely instrumental. We agree that harm prevention is important, but we suggest also scrutinizing how technologies may help to further certain positive ethical goals. Second, ethics is often operated quite abstractly and without concrete *material* for reflection to focus on. This is a problem for the task of designing specific solutions and for identifying the true ethical issues behind a certain piece of technology. Often the real value of a technical solution (as well as the true potential for harm) is only visible in the interaction with the concrete technological artifact. Our methods for ethical evaluation must be more concrete in order to tackle these issues.

With this chapter we suggest a novel approach to ethics workshops in technology that allows us to take advantage of ethics for a more productive contribution to the design process. To achieve this, we will present the concept of "virtuous practices" combining practice theory with virtue ethics. Asking whether a piece of technology is virtuous aims at focusing the evaluation of technology on its potential to facilitate practices which can be identified and justified as good from an ethical standpoint. In combining this theoretical concept with elements from well-established performative methods in design as well as theater research (see Dörrenbächer et al., 2021; 2020; Scheidegger, 2016), we seek to make ethical evaluation more concrete, beyond thinking merely abstractly about possible effects of technology.

While design research already has established concepts and methods that combine practical aspects with a wellbeing-driven approach to the design of technology (Hassenzahl, 2013; 2018; Klapperich et al., 2019) we think it is important to depart from such more *subjective* accounts in the context of ethics workshops. In this contribution, therefore, we pursue a decidedly philosophical perspective that focuses on reflecting on the conditions under which human life forms flourish. These conditions differ from individual assessments of what is positive or from feelings of happiness in their justification: something is not just good because it feels good, but because it fulfills the conditions that are constitutive for the realization of a good life as a general principle; a principle which itself needs to be justified. While psychological assessment in design for wellbeing aims at subject-related *descriptive* criteria to inspire design, *ethical reflection* must carry its evaluations beyond that and needs to ask whether those practices that are afforded by technological artifacts can be identified as good from a more general *normative* standpoint (for accounts of critical reflections on social practices, however outside of the technology contexts and not (explicitly) tied to virtue ethics, see for example Jaeggi, 2013 as well as Celikates, 2009).

# CHALLENGES OF CURRENT ETHICAL EVALUATION OF TECHNOLOGY

It is common practice to use the format of ethics workshop to integrate ethical aspects in technological development projects.
At least one ethics workshop was conducted by each of the eight robot development projects presented in this book. During such workshops, stakeholders including engineers, computer scientists, psychologists and ethicists work together to identify and mitigate ethical challenges presented by the technological solutions they are developing. Ideally, such a workshop takes place at the beginning of a project, and then again later in the process to evaluate whether challenges are being tackled appropriately and whether any new issues have emerged. Some research projects organized several workshops with different stakeholders to ensure that no important perspective is left out.

Several frameworks and models have emerged to support the planning and execution of such ethics workshops. In German technology projects funded by the *BMBF (Federal Ministry of Education and Research), MEESTAR (Model for ethical evaluation of socio-technical arrangements)* or some variant of it seems to be the most popular one. *MEESTAR* aims to assist in the evaluation of a specific technical solution for a specific person in a specific social environment with a specific need for assistance (Manzeschke, 2015). To identify possible harm that could be caused by the implementation or use of a technology being developed, *MEESTAR* provides a list of ethical dimensions that are designed to help focus the discussion around moral issues. Considering ethical dimensions such as *justice* and *autonomy*, at three perspective levels (individual, organization and societal) and four levels of "ethical sensitivity" (reaching from "application is ethically harmless" to "application is to be rejected from an ethical point of view"), the evaluation focuses on questions such as: "what are the implications of this specific piece of technology for this specific user with respect to her autonomy, and how significant are they?" From our perspective, models like *MEESTAR* represent an important step in the evolution of technology evaluation because they encourage the whole development team to engage in ethical reflections rather than delegating ethical evaluation to a dedicated group of ethicists. However, we think that there are two particular aspects of ethical evaluation where there is room for further growth and development. First, *MEESTAR* and other similar models focus too much on abstract, imaginary situations. Second, positive ethical aspects regarding the value of the piece of technology and its intended purposes are largely ignored. In the following section, we will discuss these issues, before presenting our suggestions for improvement.

# IMAGINING TECHNOLOGY USE IN THE ABSTRACT IS NOT ENOUGH

When taking part in *MEESTAR* workshops for robot development projects, more than once we found a small subgroup of the development team tasked with identifying ethical challenges. Let's consider an imaginary example of such a case. Imagine a research project that has secured funding to develop a fitness coaching robot. The robot tracks how often users exercise and has the function of motivating them to engage in more physical activity. The research team has defined a core user persona which they have named Tanja. Tanja is a 75-year-old woman who lives in a retirement home and has problems motivating herself to engage in physical exercise. The development team has also brainstormed some possible ways the robot might motivate Tanja to exercise, such as informing her of the positive effects of exercise, appearing sad when she does not exercise, or shouting like a drill sergeant. Now they bring in an ethicist to conduct an ethics workshop. How should the team respond to the question of how such a robot might affect Tanja's autonomy? Or—taking the social perspective—the question of which wider changes in society such a robot might cause with respect to the principles of justice?

In our case, when confronted with such questions, we tried to imagine situations in which the technology under investigation would have some impact on Tanja's life, other individuals and the society in which she lives, and tried to evaluate this hypothetical impact. What can be drawn from this example, and from our past experience, is that *imagination* is a critical resource for generating objects for ethical evaluation. Workshop participants need to be able to mentally simulate possible effects in order to facilitate ethical evaluation.

On one hand, the process clearly benefits from the fact that a group of people with different perspectives can identify and discuss ethical challenges, and from being structured around particular concepts like those used in *MEESTAR*. On the other hand, however, there are clear drawbacks to relying on imaginative capacities to such a great extent. More than once, it turned out during a discussion that individuals had very different mental models or imaginary representations of the piece of technology, its capacities and how to interact with it. A common understanding was missing. Evaluating something together when the *something* in question is understood differently by different people is obviously not an easy task, and often forces participants to spend a lot of time establishing a common understanding.

More importantly, relying solely on the imagination to address possible challenges often fails to produce truly new insights, distinct from what is already known. Of course, it is extremely important to make explicit what is known, and to reflect on it through joint deliberation. However, it seems unclear how such collective deliberation produces new insights into ethical implications. In short, when it comes to evaluating technology from an ethical standpoint, there is a need for more concrete forms of simulation or testing the new aspects, since we are developing an *innovation*. If designers and engineers were able to see what happens when users interact with rough prototypes designed, they could gain new perspectives beyond their subjective imaginations (for methods that enable designers and engineers to see what happens, see Dörrenbächer and Hassenzahl, 2019; Dörrenbächer et al., 2021).

# MORALITY IS NOT ENOUGH

The aim of models like *MEESTAR* is to deliberately safeguard against negative ethical effects. The best result that can be achieved in the assessment of ethical sensitivity, therefore, is *free from negative ethical effects*. Models like *MEESTAR* try to ensure that ethical *minimum requirements* are met, and to avoid unacceptable harm from the use of technology. Of course, it is vitally important to meet these minimal conditions. In all cases, avoiding harm must be the first priority. However, this exclusive focus on ethical harm can mean that ethics becomes a purely *negative* element of the engineering process. Rather than developing a positive vision of a technology, it aims only at neutralizing any potentially negative effects. As a result, the outputs of such ethical reflections are often prohibitions and restrictions, as well as regulations or sometimes smart solutions designed to prevent ethical harm. There is, however, no clear-cut distinction between restrictions and positive ethical aspects. Some moral restrictions clearly also have a positive ethical point and may be used as guidelines for technological innovation (see Misselhorn, 2022). One example is the moral principle that technologies should promote human autonomy instead of interfering with it. We suggest in this contribution to expand the focus of the positive and future-directed deliberations from moral issues in a narrow sense of the term to ethical questions regarding the good life, and how we—as individuals and as society—want to live (see also Misselhorn, 2018; 2020, where she develops a conceptual design for a software module for a care system that is capable of enhancing its users' autonomy).

Such a *positive approach* to the design of technology is also quite common in design research (see Desmet and Hassenzahl, 2012; Desmet and Pohlmeyer, 2013; Hassenzahl, 2018; Dörrenbächer and Hassenzahl, 2019). However, since these approaches focus on "happiness" and "subjective wellbeing," they differ from what is demanded by an *ethical* evaluation of technology. Ethical evaluation might take happiness or subjective wellbeing as possible sources of normative justification into account but goes further by asking whether certain objects (actions, practices, artifacts) are *good* according to an objective or at least intersubjectively binding ethical principle (which itself must undergo justification). To be clear, an ethical evaluation strives for an objective or at least intersubjective justification. Here we draw on virtue ethics as an orientation to reflect on socio-technical practices.

But how can we design technological artifacts with a constructive ethical standpoint in mind? Below, we present the theoretical building blocks of a new ethics workshop, that builds on a combination of virtue ethics, practice theory, as well as performative methods.

## COMBINING VIRTUE ETHICS WITH PRACTICE THEORY

**Virtues** are stable (character) dispositions to do (as well as think and feel) the right things in the right kind of circumstances. For example, a brave person neither recklessly charges into action nor cowardly avoids the situations but rather acts appropriately. Virtue ethicists focus more on such virtuous (character) traits when analyzing the question of what it means to be good than focusing on duties or rules. (see Hursthouse and Pettigrove, 2018).

As a first building block, we suggest extending the focus of the ethical evaluation to take into account not only moral rules, but also → virtues. This broadening of focus has the advantage that we can go beyond "moral minimalism" and ask whether a specific piece of technology is likely to have a positive impact in terms of how we want to live.

As a second theoretical building block, we suggest changing the *object of evaluation*. Although this may already implicitly be part of the models mentioned above, we suggest being more systematic and explicit about this: in an ethical evaluation, our attention should not focus solely on the piece of technology itself, but rather on the socio-technical practices that the technology enables or affords in a given context. Focusing on socio-technical practices has the advantage that these are highly concrete, and it also enables scholars to make implicit presuppositions, role-requirements and assumed know-how explicit. Instead of asking whether a specific robot is good or bad, we need to ask whether the specific practices that are afforded by the robot will have a positive impact in terms of how we want to live.

Combining these two building blocks leads to the idea of *virtuous socio-technical practices*. Virtuous practices are an ideal which we should try to achieve in the design of technology, and should serve as an anchor point for the evaluation of robots. We explain each building block below, before explicating the method itself.

## VIRTUE ETHICS

Obviously, it is not possible to consider all the details of virtue ethics in this short contribution; instead, we aim at outlining the most important and relevant aspects (for a general overview, see Hursthouse and

Pettigrove, 2018). In philosophy, it is common to distinguish virtue ethics form two other ways of thinking about ethics: consequentialism and deontology.

Consequentialism focuses on the consequences of our actions and tries to provide a set of general principles that can be used to answer the question, "what should I do?" For example, one such principle might be: "act to further the greatest pleasure for the greatest number" (for an overview of such utilitarian principles, see for example Driver, 2014). For example, in considering the question, "should I kill animals and eat them?" the point of reflection is shifted towards the question: "is the suffering and harm of animals outweighed by the pleasure you get from eating meat?"

Deontology, on the other hand, focuses on the principles that guide our actions, and provides us with rules (or tests for rules) that tell us what we must do and what we are not allowed to do. The most famous version of a self-reflective test for the principles that guide our actions is, of course, the categorical imperative: "act only in accordance with that maxim through which you can at the same time will that it become a universal law" (Kant, 1996, G 4:421). Am I allowed to lie? Well, can you imagine a world in which everybody lies? Even if some say, "of course you can!" it turns out, on reflection, that in such a world assertions of truth would be meaningless, because we could never take them as attempts to deliver a true message. The very idea of lying gets undermined, because lying is parasitic to producing meaningful communication, which itself is a presupposition of lying. This is why the whole maxim is self-defeating, and not acting according to the principle is irrational.

Virtue ethics, on the other hand, is not so much concerned with what you must do or what you are not allowed to do from a moral standpoint, but it has a different focus. The overall question of virtue ethics is: "how should I live my life in order to live a *good* life?" Virtue ethicists like Aristotle answer by advising those who wish to live a good life to espouse certain virtues. These include bravery, justice, gentleness, magnificence or, as a more contemporary example, *sustainability*. The important theoretical point here is that a concept like sustainability does not primarily prescribe how to behave, but is rather a practical orientation that makes a difference to how we lead our life. We should keep this idea in mind when it comes to evaluating technology. The suggestion is that we expand our ethical investigations by introducing the question of whether a piece of technology is good, i.e., whether it promotes certain virtues that we identify as positive in relation to the essential conditions that must be met, in order for our life forms to realize their full potential. More precisely, as we will explain in the next paragraph, this means focusing on whether the socio-technical practices that are enabled by a piece of technology are virtuous, i.e., whether they can be justified as contributing to the flourishing of (human and other species) life forms.

# SOCIAL PRACTICES AND SOCIO-TECHNICAL PRACTICES

So far, we have suggested that designers, engineers and ethicists should not only ask whether *it* is allowed, but also whether *it* is good. But what is *it*? It might be clear—at least implicitly—that ethical evaluation cannot consider a piece of technology in isolation, or in a socio-technical vacuum. It is not the abstract piece of technology in itself that is the object of evaluation, but rather the way it enables and encourages us to interact with it. We thus believe that it is beneficial to make this pattern of interaction the explicit object of evaluation. While virtue ethics was the theoretical building block supporting the extension of evaluation criteria from moral minimalism to the question of goodness, practice theory is the theoretical framework that carries the second shift from technological *objects* to socio-technical practices.

What are → social practices, and how can a theory of social practices help to improve the ethical evaluation of technology? A *practice-theoretical* perspective on sociality and society has been proposed by several authors (e.g., Giddens, 1984; Bourdieu, 1979; Schatzki, 2008, to name a few). In essence, social practices are forms or ways of acting and interacting that demand a certain amount of normative sensitivity (i.e., they presuppose knowledge of implicit or explicit rules) from individuals who intend to participate in a practice. There are, for example, specific ways to greet people in different social contexts. If you do not follow the patterns of behavior that people expect you to follow, you may be criticized or sanctioned for greeting in an unusual, unexpected or impolite way.

**Social Practices** are socially reproduced forms of interaction. Sometimes such forms or ways of interaction can be made explicit by referring to rules or rule-like terms. Socio-technical practices are social practices in which technology plays an essential mediating or even agential role.

A key insight highlighted by Giddens (1984) is that such practices are simultaneously enabling and limiting. If you want to express yourself in a given language, you are required to do so according to the rules and structure of that language, and these rules and structure constrain your attempts to express yourself. At the same time, it is only because such a language structure exists that you are able to express yourself at all. This is how certain practices both enable and create a horizon for possible actions, while also demanding that specific actions take a specific form.

How does technology fit into all of this? Socio-technical practices are interesting in two ways. First, a certain piece of technology, a social robot, an ambient home assistant or a smartphone, can become an essential part of a social practice, and enable specific forms of action and interaction. The *rules* of how to act and interact with a given piece of technology are embodied into its design, function and shape, and demand specific actions from us while inhibiting others. Second, technology can transform patterns of action and interaction, or even create entirely new ones. This latter case is interesting because it emphasizes the fact that creating technology is not only about creating objects, but rather about creating socio-technical practices, i.e., new forms of interaction. To introduce a slogan: designers and engineers are creators of socio-technical practices (see also Feige, 2018).

# VIRTUOUS PRACTICES

Now that we have introduced these individual components, we can move on to synthesizing them into a concept of *virtuous* practices.

By drawing on virtue ethics, we extend ethical evaluation beyond moral considerations. At the same time, virtue ethics ensures that we do not settle the question of whether a piece of technology is good by only focusing on subjective wellbeing, but by asking whether a piece of technology (and the practices it affords) leads to ethically desirable social practices which form together a certain life form. The aim is to identify the conditions under which life forms flourish, i.e., conditions that are constitutive for the realization of a good life according to, and justified by, general and justifiable principles. A very generic way of putting this is to say: a good life form is met when practices are able to concretely increase individual agency while at the same time making sure that agency is intersubjectively shared and realized.

By drawing on practice theory, we focus our investigation on the socio-technical practices within which a specific piece of technology is embedded, and which it enables or affords.

In this way, our ethical evaluation aims to consider whether the specific socio-technical practices afforded by a specific piece of technology are good.

Further reflection then allows the specific nature of this ethical goodness to be expressed through identifying and explicating the virtues of specific socio-technical practices. After we have identified a certain practice as good, we can then by way of *reverse engineering* carve out why it is good, and what type of virtues it embodies.

But how can we identify whether a socio-technical practice is in fact good? This is where our last shift comes in: we suggest a novel approach for ethics workshops which combines the theoretical building blocks described above with elements of performative methods that were developed in design research (Dörrenbächer et al., 2020; 2021; (→ p. 140) as well as in theater research (Scheidegger, 2016).

CATRIN MISSELHORN, MANUEL SCHEIDEGGER, TOBIAS STÖRZINGER

# COMBINING VIRTUOUS PRACTICES WITH PERFORMATIVE METHODS—PLAY HARD, REFLECT EASY

The challenge for everyone involved in designing new socio-technical practices is that we often have a specific (and rather abstract) vision of how a piece of technology should be used, and how interacting with it will affect the lives of users and society in general. However, whether practices are ultimately good depends on whether such interactions genuinely lead to a sustainable positive impact on the users' lives, and contribute positively to how we want to live. How can we find out whether the practices that we create really have such virtue, i.e., whether they really contribute to the flourishing of human life forms?

In order to solve this problem, we propose combining the theoretical concept of virtuous practices with elements of established performative methods (see above).

The potential of combining these theoretical concepts with performative methods lies in the fact that the virtual acting out of practices allows to provide material and first entry points for an ethical reflection:

> In virtual actions (i.e., theatrical situations) we [...] recognize relations, appropriateness, indeterminacies of behavior, feeling, social being and bodily reacting in a concrete experienced situation of acting, which we can relate to our usual acting. This is an essential mode of reflection: to understand the concreteness of an experienced situation as a regulative idea of one's own action (Scheidegger 2016, p. 56, our translation).

Performative methods use rapid prototyping (a method that is also common in the "Design Thinking" approach) and theatrical role play to generate insights on how practices that are enabled by a piece of technology unfold. However, it is important to make clear that the goal of using these methods is not user acceptance testing for products that will soon go to market. Instead, they should be used to create knowledge about the fundamental presuppositions behind our artifact, and even ourselves. Rapidly building usable prototypes and role-playing with them allow us to go beyond pure imagination when it comes to ethical evaluation. As we have indicated above, prototyping, role-playing and focus groups with different stakeholders are an established method in design research, theater research and other design methods such as Design Thinking. Using these methods in the context of ethics of technology as well as ethics workshops allows us to get a better understanding of our object of evaluation.

Overall ethical evaluation can then make use of simulating and testing ethical implications in concrete situations by taking the results and material for asking ethical questions of whether the simulated practice that is enabled by the fictional artifact can be identified as good. Does it contribute to the way we want to live and contribute to the flourishing of (our) life form(s)?

Thus, enacting and being *playable* is an essential element which sets this kind of method apart from more abstract methods like *MEESTAR*. Instead of merely imagining the possible effects of a certain robot, we can simulate them by playing through the socio-technical practices it affords and test several implications this way.

Interesting questions for reflection might be: how do children, older people, people of different genders, or people with disabilities interact with the prototype? Do our presuppositions of how to use it turn out to be correct? Do users benefit from using it? What impact does it have on society? And, especially, does it reliably lead to the realization of those (justified and reflected) values that we have associated with it in the first place?

In role-playing through a practice, we might, for example, realize that we assumed that all users of the robot would be able to read. We might learn that users initially enjoy interacting with the robot, but soon lose interest in it. It could turn out that a certain interaction is not good because it is simply not realizing any virtuous practices that we originally thought it would.

If enactments from different perspectives are successful, we should gain more concrete data for evaluation and reflection than we would have obtained from merely imagining possible use scenarios. Concrete interactions enable us to reflect on whether the playthroughs really showed the virtues we imagined when designing the piece of technology, and whether the piece of technology and the practices it affords really add to the lives of users and to society as a whole.

Next, we suggest a further step for simulating and testing the socio-technical practices we have designed. Since it is not always possible to empathically take on the perspectives of others, such testing and simulation will benefit greatly from engaging a diverse range of users and stakeholders (Misselhorn, 2020, calls this a new type of "Experimental Philosophy"). Ideally, it is not just the design team or the engineers who will be involved in testing and reflecting on ethical virtues, but also potential users themselves. Involving potential users and stakeholders in the design process by integrating their feedback on their interactions is an essential part of answering the question of whether a new piece of technology can be considered good in general.

The three key shifts we suggest can be summarized as follows:

1. Object of evaluation: take into account not just an artifact existing in a vacuum, but rather the socio-technical practices it produces, i.e., how users interact with the designed artifact.

2. Evaluation criteria: reflect on whether those socio-technical practices are good in a wider ethical sense. Do they contribute positively to the good life of users, and the way we want to live as a society? Reflect not only on whether the practices enabled by the new technology are morally permissible, but also on whether they actively embody particular virtues.

3. Use pre-designed and prototyped practices in role-play and enactment in order to simulate or test designed practices. Ideally, test these with different users and stakeholders, and consider a broad range of perspectives in your reflections on whether the practice can be considered good.

# PRELIMINARY PRACTICAL SUGGESTIONS AND THE WAY AHEAD

The theoretical model we have presented here, as well as the ethical evaluation method for technological development processes itself (→ Fig. 1), must be tested, iterated and refined. We are only in the early stages of creating and designing this (meta-)socio-technological practice, incorporating extensive feedback we have received from designers, engineers and ethicists. Let us conclude by providing an initial process outline that we have already successfully implemented in the form of a one-day workshop.

We suggest dividing the process into three different phases, each consisting of three subphases:

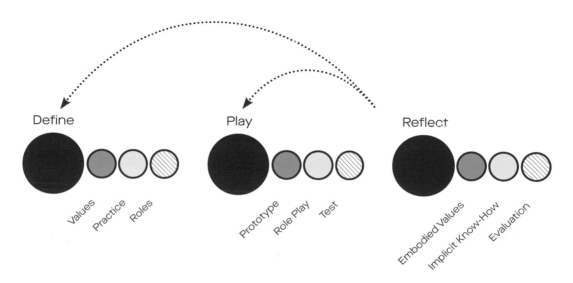

CATRIN MISSELHORN, MANUEL SCHEIDEGGER, TOBIAS STÖRZINGER

**Fig. 1**  Phases of our workshop prototype.

Before we explain each phase in more detail, let us mention that it is possible to precede the define phase with an ideation phase, and therefore to use this workshop in the very early phases of development. This is particularly useful if the aim is not yet an ethical evaluation of an already fully defined artifact, but to develop and ethically evaluate ideas for technical artifacts in the first place. In our first workshop with participants from different fields of robot development, we therefore started by letting groups of participants identify a specific practice that they judged to be problematic. Second, they reflected on *why* this practice is problematic and then determined which ethical values could be realized by improving the practice. Subsequently, technical solutions were brainstormed and fleshed out. These product ideas were then used to start the define phase as described below.

# THE DEFINE PHASE

The define phase should function as a theoretical frame, which will prepare the participants mentally and serve as a guide for the practical activities involved in the subsequent stages. However, it is also important not to overtheorize at this stage, as the aim is to develop an initial shared understanding of ethical values, the socio-technical practices that are afforded by the piece of technology, and the roles involved. The first subphase *(values)* should develop an initial basic understanding of how the envisioned piece of technology should contribute to a good life for its users. Which dangers or moral prohibitions need to be considered? Which ethical values underpin the technology? The second subphase *(practices)* aims at making explicit the socio-technical practices that are enabled or afforded by the piece of technology. It is helpful to write down concrete examples of how users might interact with the artifact and then discuss the implicit knowledge that is presupposed by this interaction. The third subphase *(roles)* focuses on the different roles that are involved in socio-technical practice. What roles are required in order to keep the interaction stable, and what are the requirements to fulfill these specific roles?

# THE PLAY/ENACT PHASE

The main function of the enactment phase is creating concrete objects for reflection, as well as gaining insights into how interaction with the not-yet-existing piece of technology turns out. This phase can be thought of as simulation or testing of the envisioned artifact. In order to reach this goal, the first subphase *(prototype)* starts with creating a prototype of the socio-technical practice. This means first prototyping the artifact itself. This may involve finding creative solutions for rapid prototyping from Design Thinking. It is not usually possible to build a robot in 15 minutes, so one simple solution might be to put someone in a box and ask them to act like a robot (although of course there is still the possibility that this will lead to distortions, as there still is a person involved). Second, it is important to define who plays which role, and what the rules are for playing this role, as well as the more general restrictions involved. The second subphase *(role play)* consists in playing through the design practice. We suggest including several enactments, switching roles and considering the perspectives of various different stakeholders. Observers should note down what is going on. In the third subphase *(test)*, we move from role-playing to pre-testing with external stakeholders, allowing them to explore how their interaction with the artifact unfolds in practice. In this book you find similar role-play approaches (→ p. 114, p. 140).

# THE REFLECTION PHASE

The main aim of the reflection phase is to answer the question of whether the technology is good. Does the socio-technical practice afforded by our prototyped piece of technology contribute to the good life of users and the way we want to live as a society? Are ethical dangers and worries founded? Instead of addressing this question in an imaginary vacuum, we can use what we have learned from simulation and testing as concrete examples for evaluation. The first subphase *(embodied values)* consists in reflection and deliberation on the values that have been expressed through practical simulation or testing. What is the true nature of the practice, and which values does it embody? Were our assumptions of how the piece of technology would contribute to the user's life correct, or did it turn out to embody other virtues, or even vices? The second subphase *(implicit know-how)* involves reflecting on the implicit knowledge that users who will interact with the technology are assumed to have. This is important because it might turn out that we have gained new insights about our own presuppositions when creating the technology and designing the interaction. We might find new requirements for a successful implementation of our envisioned values at this stage. The third subphase *(evaluation)* concludes the process with an overall ethical evaluation. We suggest taking the embodied values, virtues and know-how into account when evaluating the practice as a whole, and reflecting on possible violations of basic moral principles. This overall evaluation also leads to two possible iterative steps. First, we can go back to the enactment stage and repeat this process with changes to the prototype and/or the roles, or introducing new perspectives. Alternatively, we can go back and restart the process from the beginning after redesigning the core values of our design project. Ideally, this process should introduce a form of learning to the process, allowing it to gradually become more virtuous.

## THE ROAD AHEAD

This method itself needs to be further tested and amended for optimal implementation. However, we are confident that the theoretical concept, as well as the suggested workshop prototype, is a step in the right direction. We believe that enacting the practices enabled by new technology will lead to improvements in the ethical evaluation process, and that designers, engineers and ethicists must move beyond asking whether something is allowed, and instead ask whether it is good.

We would like to thank Marc Hassenzahl and Judith Dörrenbächer for the discussion and their many insightful comments which helped us relate and distinguish our philosophical approach to established approaches in the design research such as their approaches called "Positive Futures" or "Enacting Utopia."

Bourdieu, P. (1979). *Entwurf einer Theorie der Praxis: Auf der ethnologischen Grundlage der kabylischen Gesellschaft*. Suhrkamp Verlag.

Celikates, R. (2009). *Kritik als soziale Praxis: Gesellschaftliche Selbstverständigung und kritische Theorie*. Campus Verlag.

Desmet, P.M., & Hassenzahl, H. (2012). *Towards happiness: possibility-driven design*. In Iba Zacarias, M.; & Valente de Oliveira, J. (Eds.), *Human-computer interaction: The agency perspective*, Springer, 3–27.

Desmet, P.M. & Pohlmeyer, A.E. (2013). Positive design: An introduction to design for subjective well-being. *International Journal of Design*, 7(3), 5–19.

Dörrenbächer, J., & Hassenzahl, M. (2019). Changing perspective: A co-design approach to explore future possibilities of divergent hearing. In *Proceedings of the 2019 CHI Conference on Human Factors in Computing Systems*, 1-12.

Dörrenbächer, J., Laschke, M., Löffler, D., Ringfort, R., Großkopp, S., & Hassenzahl, M. (2020). Experiencing utopia. A positive approach to design fiction. *Workshoppaper Submitted for CHI'20*.

Dörrenbächer, J., Laschke, M., & Hassenzahl, M. (2021). Utopien erleben. Eine Methode für soziale Innovationen aus dem Jahr 2020. In Röhl, A., Schütte, A., Knobloch, P., Hornäk, S., Henning, S., & Katharina, G. (Eds.), *Bauhaus-Paradigmen. Künste, Design und Pädagogik*. (pp. 369–377). De Gruyter.

Driver, J. (2014). The history of utilitarianism. In Zalta, E. N. (ed.), *The Stanford Encyclopedia of Philosophy (Winter 2014)*. Metaphysics Research Lab, Stanford University. https://plato.stanford.edu/archives/win2014/entries/utilitarianism-history/

Feige, D. M. (2018). *Design: Eine philosophische Analyse* (Originalausgabe Edition). Suhrkamp Verlag.

Giddens, A. (1984). *The constitution of society*. University of California Press.

Grunwald, A. (1999). Technology assessment or ethics of technology? *Ethical Perspectives*, 2, 170–182.

Grunwald, A. (ed.). (2013). *Handbuch Technikethik*. Verlag J.B. Metzler.

Hassenzahl, M., Eckoldt, K., Diefenbach, S., Laschke, M., Lenz, E., & Joonhwan K. (2013). Designing moments of meaning and pleasure. Experience design and happiness. *International Journal of Design* 7(3), 21–31.

Hassenzahl, M. (2018). The Thing and I (Summer of '17 Remix). In M. Blythe & A. Monk (Eds.), *Funology 2: From Usability to Enjoyment* (pp. 17–31). Springer International Publishing.

Hursthouse, R., & Pettigrove, G. (2018). Virtue ethics. In Zalta, E. N. (ed.), *The Stanford Encyclopedia of Philosophy (Winter 2018)*. Metaphysics Research Lab, Stanford University. https://plato.stanford.edu/archives/win2018/entries/ethics-virtue/

Jaeggi, R. (2013). *Kritik von Lebensformen*. Suhrkamp Verlag.

Kant, I. (1996). Groundwork of the metaphysics of morals. In Immanuel Kant: Practical philosophy (M. J. Gregor, ed.; translated by Wood, A. W.). Cambridge University Press. (Original work published 1785).

Manzeschke, A. (2015). MEESTAR: Ein Modell angewandter Ethik im Bereich assistiver Technologien. In Weber, K., *Technisierung des Alltags—Beitrag für ein gutes Leben?* 247–262.

Mickle, T. (2021) Google plans to double AI ethics research staff. *Wall Street Journal*. https://www.wsj.com/articles/google-plans-to-double-ai-ethics-research-staff-11620749048

Misselhorn, C. (2018). Grundfragen der Maschinenethik (4., durchges. Aufl. 2020). Reclam, Philipp, jun. GmbH, Verlag.

Misselhorn, C. (2020). Artificial systems with moral capacities? A research design and its implementation in a geriatric care system. *Artificial Intelligence*, 278. 103179.

Misselhorn, C. (2021). *Künstliche Intelligenz und Empathie. Vom Leben mit Emotionserkennung, Sexrobotern & Co.* Reclam-Verlag.

Misselhorn, C. (2022). Artificial moral agents. Conceptual issues and ethical controversies. In Vöneky, S., Kellmeyer, P., Müller, O., & Burgard, W. (Eds.), *The Cambridge Handbook of Responsible Artificial Intelligence: Interdisciplinary Perspectives*. Cambridge University Press.

Schatzki, T. R. (2008). *Social practices: A Wittgensteinian approach to human activity and the social.* Cambridge University Press.

Scheidegger, M. (2016): Virtuelle Handlungen, reale Konsequenzen. Über Theatralität und die ästhetische Differenz des Digitalen. In: Innokentij Kreknin u. Chantal Marquardt (ed.): *Das digitalisierte Subjekt. Grenzbereiche zwischen Fiktion und Alltagswirklichkeit. Sonderausgabe #1 von Textpraxis. Digitales Journal für Philologie* (2.), 37–60.

# Elke Buttgereit

is a Vice-Principal at an elementary school in Bonn. She studied elementary education in Cologne, specializing in German, math and general studies (social studies). She has worked as a Primary School Teacher for more than 30 years. She has been grappling with the importance of robotics in schools since 2018.

# Are Robots Good at Everything? A Robot in an Elementary School

Elke Buttgereit

Elementary school children are fascinated by robots. The school I teach at is located in a socially disadvantaged area in the city of Bonn. Only very few of the children have ever encountered a robot before, but they know about them from television, books or video games. For them, robots are perfectly functioning machines that do exactly what humans tell them to, they immediately spring into action whenever they are needed, they can answer every question they are asked, they are ready to go at any time and they help in any situation.

In the fall of 2019, I had the great fortune of being allowed to *babysit Pepper*, a humanoid robot from the *University of Siegen*, for two weeks during school hours. When I introduced *Pepper* to my fourth-grade students, the children were immediately taken by it—because of *Pepper*'s friendly voice, googly eyes, human-like appearance and child-appropriate size. The first lessons with the robot were characterized by joyful anticipation, excitement and fascination. They cautiously tried to explore *Pepper*, carefully touched its hands, stroked its head, and felt validated when *Pepper* made noises full of joy.

The guidelines and curricula of every German state do not explicitly require students to explore the subject of robotics. It is generally formulated that students should learn to "handle media responsibly and safely" for the tasks and objectives at hand (cf. Section 2 (6) and (5) SchulG). Media literacy is most particularly understood as relating to the use of new media to research and present information, and to use it in a meaningful way. In my opinion, however, direct engagement with robotics, as part of a more comprehensive education in media, is essential in the digital world that children are growing up in today. Robots have been providing us with support for more than 60 years, but this mostly remains hidden from the eyes of children.

The robots work in industries and places humans cannot reach, on Mars or in the depths of the ocean. Nevertheless, they also help us at home, with vacuuming or mowing the lawn, or they give us directions, like at the airport or in a museum. Self-driving cars are being put to practical tests, and robots have become an indispensable part of our lives. They are becoming increasingly intelligent, more human-like, and are very likely to become part of our future. This raises key questions that children should be addressing in school. What are the opportunities and risks for our future? And what about the ethical concerns?

I designed a series of lessons around *Pepper's* visit. We tackled what a robot actually is (a programmed machine), where they are used (at home, in industries or even as movie stars, for instance), and why. The children learned the technical components of robots and even built their own small cleaning robot using a toothbrush, a vibration motor, coin cell batteries, and wobbly eyes, amongst other things. We discussed how the *Turing Test* is used to distinguish between human and machine, and the meaning of Isaac Asimov's *Three Laws of Robotics* (1942). The lessons also addressed the historical development of robotics, and we speculated about the possible future of robots.

I carefully prepared ethical questions that were then expanded in conversations with the children, where they expressed their thoughts and reflections. The following questions accompanied us throughout the series of lessons:

* Should robots look like humans and become increasingly more human? Will this improve the way we live together? Can you imagine talking to a robot as if it were a human being?

* Can a robot become a real friend, or even your best friend? What will happen to your human friends?

* Can there be rights for robots like there are children's rights? Are robots allowed to have free time?

* Robots have cameras built into them—could they perform supervision? As supervisors in the playground at school, in the classroom, on the street, in your room at home?

* Can robots become more intelligent than humans? What will happen if one day robots are smarter than your parents, police officers, doctors, scientists and politicians? What would be good about this and what would be bad?

* Would you like to be taught by robots? How do you imagine class when a machine explains parts of speech or talks to you about friendships? (This series of questions was something that really engaged my fourth graders. While the children were mostly in favor of robot teachers at first, in the end they were more in favor of human teachers, since they "also make mistakes, can laugh, and are not as monotone" as robots. "But it's something I could imagine for mental arithmetic quizzes, especially for different levels. As a reward, I would be allowed to play another game.")

* Today, robots are working in retirement homes and provide support for the care staff. Would you prefer visiting your grandparents in a retirement home if there were robots there? Do you want your grandparents to spend time with a robot?

While we were philosophizing, it was important to me that I not take on the role of the *omniscient teacher*, but that I looked for possible answers on an equal footing with the children. The questions here cannot be answered with a clear yes or no, but we can make children more sensitive to the issues. That is why I repeatedly worked to distinguish between human and machine with the children. It became clear that a robot can only do what it was programmed to do; it is only as good as its programmer.

While the children had high expectations from robots at the beginning of my series of lessons, they gradually realized that robots cannot perform as many tasks as they initially thought. They are not perfect. And *Pepper* is not, either. Their patience was tried even by the long stretch of time between turning it on and being able to use it. *Pepper* cannot do their homework for them, cannot clean up, play catch outside, go swimming, or show real emotions. Robots are not good at everything, despite numerous media outlets suggesting such.

But will there be a *s/he* one day, a robot with consciousness, and, if so, what will the consequences be? This is a question that clearly needs to be asked, even in elementary school.

Now the way needs to be paved for new technologies to remain governable. Are robots good at everything? Children need to become more sensitive to this issue. Whenever I am asked whether robotics is really an appropriate subject for our elementary school students, I say: these children will live with robots later on; they are the ones who will be most affected by this development—it is essential for this subject to be taught in elementary schools.

Asimov, I. (1942). Runaround. *In Astonishing Science Fiction*. Street and Smith.

## Scarlet Schaffrath (Siebert)

is a communication scientist and PhD student at *TH Köln—University of Applied Sciences*, Institute for Media Research and Media Education, and analyzes future visions of social robots in educational contexts. In her research she focuses on participatory approaches to technical innovation. She also works as a scientific consultant at *VDI/VDE-IT GmbH*.

# The Medium has a Message—Educational Robots in a Didactic Triangle

Scarlet Schaffrath

Whenever I talk about the *MERITS* project, in which my colleagues from linguistics, media education and I research the potential of social robots for educational contexts, I get the same reaction: amazement. "What do robots have to do in kindergarten?", "Do you really want to replace human teachers?", or "Is the technology ready yet to teach languages to children?" are questions frequently asked by friends and family. In the scientific context, too, I have to justify our research interests. Depending on the discipline, the discussions revolve around the anthropomorphization of robots, the usefulness of robots for learning effects, or the necessity of digital skills in the 21st century.

* The humanities ask questions such as: Why do robots look so human-like? Couldn't they be shaped completely differently, making it more obvious that they are technological artifacts rather than living beings? Perhaps, then, children would not so easily anthropomorphize and misconstrue them.

* Psychology argues: Why don't you do experimental studies in which you compare humans with robots? Then we would finally have proof that robots are helpful tools for learning.

* Pedagogy demands: Shouldn't children learn to deal with new technologies in terms of being able to program them instead of treating them like social counterparts? Skills in Science, Technology, Engineering and Mathematics (STEM) are becoming increasingly important in the digital age.

All of these questions are of interest to me as a communication scientist, because they point to one thing: a different standard is applied to social robots than to *classic* digital technologies or media;

humans seem to serve as a comparative dimension for them. This is particularly evident in expectations of how we should communicate with robots; they should be able to *speak* naturally *with* humans using verbal language and gestures, rather than only transmitting information *between* people as traditional media was expected to do.

However, this seemingly natural interaction does not always work, as I observed during various child-robot interactions, both in everyday environments such as libraries, and in scripted situations, such as laboratory settings. In both circumstances, children were usually extremely patient: although robots demonstrate learning-supportive gestures only in slow motion, and barely *understand* 30 to 40 percent of children's communication attempts, children ask their questions (e.g., what's your name? How old are you?) repeatedly, getting slower and louder on each attempt.

Adults contribute to children's patience towards robots. Technology developers, as well as educational specialists and parents, anthropomorphize robots with explanations such as, "the robot is tired" (instead of "the battery is empty") or "*NAO* is shy" (instead of "the robot's speech recognition is still developing; it did not process your request"). These explanations conceal the fact that robots are technical artifacts that struggle with natural interaction. Rather, they promote the perception of robots being social counterparts with emotions and desires, which then leads to more social behavior and better learning outcomes, respectively. For example, children are more likely to listen to embodied social robots than to virtual agents, and to follow their instructions. Therefore, robots are also seen as a useful tool outside of STEM education.

As a communication scientist, this anthropomorphization is interesting insofar as when technology becomes a social counterpart, "the medium is the message" (Marshal McLuhan) turns into "the medium has a message."

Against the background of the vulnerability of young learners and the technological inexperience of educational specialists, consideration should be given to how the opportunities presented by new and supposedly social technologies can be used, and how risks can be avoided or at least conveyed. Let's imagine a future scenario (similar to the scenarios in the *ERIK* project) (→ p. 58)—involving a didactic triangle of a student, a teacher and a robot:

> *At school, a teacher notices that one of her students has been lagging behind for a while. He seems absent, and she could only grade his last test with a D. The teacher consults the robot, which has been sitting in the classroom accompanying the class for a few weeks. Using its sensors, the robot observes the (social) behavior of the children and evaluates their attention spans, messages and disturbances (e.g., talking to their neighbors).*

*The teacher retrieves the information about the student and learns that he only raised his hand once in the last three weeks, and that on average he looks out of the window 60 percent of the time during class.*

What could the teacher do with this information? One possibility could be to offer the student one-on-one tutoring with the robot so that he can repeat the lessons. However, this requires data literacy, i.e., the competence to classify the data collected and evaluated about the student by the robot, and to be able to act accordingly. Furthermore, the robot cannot answer the question of why the student looks out of the window for most of the class.

In my opinion, interaction strategies for robots should be contextualized even more. Furthermore, new interaction strategies for human counterparts should be considered, especially when addressing children. To me, it also seems important to assess which functions and roles robots can take over in which cases. Finally, it needs to be made clear that robots may alleviate symptoms (here: improving students' grades by providing more practice), but not the actual cause of the problem (here: a student's troubles at home and their lack of concentration in regular lessons). Interdisciplinary approaches, like collaboration between the above-mentioned humanities, psychology and pedagogy, should therefore be part of the design process of human-robot interaction. Communication science might serve as a mediator between these disciplines, as well as between research and practice. This makes it possible to analyze interactional changes between students and teachers—be they human or robotic—and the learning content within the didactic triangle.

The *MERITS* project is supported by the *Digital Society research program* funded by the *Ministry of Culture and Science* of the German State of North Rhine-Westphalia.

## Dr. Ilona Nord

is Professor of Religious Education and Practical Theology at the *University of Würzburg*, Germany. Her research focuses on religion and media, digital literacy, and digital religion. She also works on diversity management in cultural contexts, antisemitism and religious pluralism. Her latest publications are the compendium *Theologie und Digitalität* (Herder 2021, co-edited with W. Beck & J. Valentin) and *Churches Online in Times of Corona (CONTOC): First Results* (Campbell 2021, with O. Adam).

# The Friendly Siblings of Workhorses and Killer Robots—Becoming Alive Through the Nonliving, and Feeling Blessed by a Religious Machine

Ilona Nord

VIVA—a social robot presented in this book (→ p. 78)—is supposed to act as a maid in private rooms, or in residential homes for senior citizens. It helps to fight against the feeling of loneliness, or at least that is the plan. That sounds good, if a bit disconcerting. After all, an effective remedy for loneliness is human togetherness, which is now being replaced by robotics. But it has long been clear that the use of robotics in social relationships is too banal to be treated as a substitute. In addition, however, the name VIVA surprises me—it is an imperative: live! But who actually calls whom into being alive?

In many European films and books, it was humans who shouted live! to robots. This is exemplified not least by an archetype of the artificial human being, the Golem of Rabbi Löw (Wöll, 2001). And then the creature—whether made out of clay (Golem) or metal (robots in science fiction)—sets itself free and becomes independent (how could it be otherwise?). The robot turns against its creator ... I suspect that the HCI community is not always aware of the cultural ambivalence associated with the construction of artificial life and humanoid robots. However, indications of this are provided by a small linguistic analysis of the English-language web, focusing primarily on two varieties of robot: "workhorses" and "killer machines" (Hurtienne, 2020). In the context of this book, robots are (or at least could be) designed as everyday friends, which—it can be reasonably assumed—is not consistent with current public perceptions. But that will change rapidly. This has already been demonstrated by *VIVA*, the robot that inverts the imperative by suddenly calling out to humans to *live!* The robot, your friend and coach, your constant companion.

The need for stability, constancy and immutability in a world of singularities that has become fluid makes intelligent entertainment partners attractive to lonely people. They keep people interacting artificially, so to speak, in forms of AI, so that no one completely falls back into their inner world and loneliness. They set impulses so that people are challenged to trust in life.

It is clear that robots will also become recognized partners in social interactions in Europe, as seems to have long been the case in Japan, China and elsewhere. This might also be a cultural change: with social robotics, peace is somehow returning to the description of the relationship between man and machine. The robot is no longer exploited as a workhorse, and its image as a fighting machine—as a killer—recedes into the background. *VIVA*, long live the friendly robot, the friend who wants to do you good, and keeps you in beneficial bonds!

*VIVA* reminds me of another robot that is a more unique religious artifact than a functional mass-produced robot. Its name is *BlessU2*, the blessing robot. The robot delivered blessings to more than 10,000 visitors to a public exhibition on the 500th anniversary of the Reformation in Wittenberg, Germany, and it continues to *work* in many places, in churches and beyond. Initially, we analyzed visitors' comments regarding content and affect. In a second study, we used *BlessU2* and another humanoid robot, *QT,* to investigate how basic design aspects, like size, gesture, voice, and facial expressions, influence the emotional experience and cognitive judgment of being blessed through robot technology (Löffler et al., 2021).

As part of the project for this book, we decided to send *BlessU2* to the picture-world of the messianic kingdom depicted in *Peace* by William Strutt (1825–1915) (→ Fig. 5, p. 35). Strutt's painting *Peace. A reign of love, peace on earth* (1903) has been popular for more than a hundred years, and copies hang in many living rooms and bedrooms around the world, including my parents' bedroom, to this day. The painting is a modern depiction of the Christian vision of the Kingdom of God. It refers to the Jewish prophet's word of the Messianic Kingdom of Peace, as portrayed in the Hebrew Bible and the Old Testament in Chapter 11 of the Book of Isaiah (v1–12, quoted from the American Standard Version, vv. 1–2 + 5–6): "**1** There shall come forth a shoot from the stump of Jesse, and a branch from his roots shall bear fruit. **2** And the Spirit of the LORD shall rest upon him, the Spirit of wisdom and understanding, the Spirit of counsel and might, the Spirit of knowledge and the fear of the LORD. **5** Righteousness shall be the belt of his waist, and faithfulness the belt of his loins. **6** The wolf shall dwell with the lamb, and the leopard shall lie down with the young goat, and the calf and the lion and the fattened calf together; and a little child shall lead them."

Our robot *BlessU2* immerses itself in the ancient myth, becomes part of the vision of an eternal life in which violence and injustice no longer have power. *BlessU2's* contribution to this suddenly becomes clear, because it stages and communicates violence interruption. A *blessing* is the primal gesture for this; to bless means "to speak well" (Gen. 1). Those who bless imitate, as it were, God's creative action on the seventh day. To love life, to make visible the good in life, even in all adversity.

The great importance of this primal gesture for many religions is made clear not least by the words: "The shortest definition of religion is interruption" (Metz, 1991). *BlessU2* was able to stage such an interruption in a humorous way, and partially disrupted expected patterns of religious practice (Löffler et al., 2021). These disruptions, however, gave rise to new questions about the relationship between Christian religion and robotics, which are equally interesting and far-reaching for religious practice, its reflection, and theology. *BlessU2's* eyes, for example, were frequently commented on by users, who said they were not vivid enough. In the world of android robots, the eyes are always considered a sign that there is an unbridgeable difference between humans and robots. That difference is called consciousness and experience, but somehow this argumentation doesn't quite work in this instance. *BlessU2* has awkwardly automatic facial expressions and looks empty. It is also by no means intelligent enough to make small talk. Even so, several thousand people have had it give them a blessing. *BlessU2* is nothing more than a metal case fed with a minimally intelligent algorithm, which is a far cry from what robotics can do today. But the discussion it sparked had worldwide impact, and is still reverberating, provoking questions and motivating people to think about religious robots and speculate about and experiment with appropriate designs (cf. Universität Würzburg, 2021).

The time of the demonization of android robots is behind us; robots do not only exist as killer robots in science fiction; their siblings, the social robots, also exist. Culturally, for some time now we have been in a romanticizing phase in which the emotional attachment to social robots and their responsible design seems particularly interesting, e.g., in the film world (cf. Worschech, 2021). I am looking forward to the time when the initial hype and irritations are gone. What kind of thoughts and feelings will take place then? But one thing already concerns me now: what will become of the faithful companions that have been a counterpart to a person for years, basically *witnessing* a large part of his or her thoughts and feelings? Then Mark Coeckelbergh's (2020) question will come to a head: "Is an AI 'just a machine,' or does it deserve some form of moral consideration? Should we treat it differently than, say, a toaster or a washing machine? ... This question is not about the ethics by or in AI but about our ethics toward AI. Here the AI is object of ethical concern, rather than a potential ethical agent itself." (49 f.) In any case, *BlessU2* has already tested its place in the messianic kingdom of peace.

Coeckelbergh, M. (2020). *AI ethics*. MIT Press.

Hurtienne, J. (2020). *Common conceptions of robots on the English web—A corpus linguistic analysis*. Workshop Metaphors for Human-Robot Interaction at the 12th International Conference on Social Robotics (ICSR 2020).

Löffler, D., Hurtienne, J., & Nord, I. (2021). Blessing robot *BlessU2*: A discursive design study to understand the implications of social robots in religious contexts. *International Journal of Social Robotics*, 13, 569–586. https://doi.org/10.1007/s12369-019-00558-3

Metz, J. B. (1991). *Glaube in Geschichte und Gesellschaft: Studien zu einer praktischen Fundamentaltheologie*. Matthias-Grünewald-Verlag.

Universität Würzburg (2021). CoTeach—Connected teacher education. https://www.uni-wuerzburg.de/pse/forschen/coteach-connected-teacher-education/

Worschech, R. (2021). Kritik zu Ich bin dein Mensch. [Review of the film *Ich bin dein Mensch*]. *epd-film*, 71. https://www.epd-film.de/filmkritiken/ich-bin-dein-mensch

Wöll, A. (2001). Der Golem: Kommt der erste künstliche Mensch und Robote aus Prag? In M. Nekula, W. Koschmal, & J. Rogall, *Deutsch und Tschechen: Geschichte, Kultur, Politik*, 235–245. C.H.Beck.

THE FRIENDLY S BLINGS OF WORKHORSES AND KILLER ROBOTS

ILONA NORD

# Appendix

## Meike Hardt

is a Designer and Design Researcher interested in the political dimension of design. She is co-founder of the *Research Institute of Botanical Linguistics (RIBL)* and worked for the project *Critical by Design?* at the *Institute of Experimental Design and Media Cultures (IXDM), FHNW Basel* as a researcher and coordinator.

# Designing with Algorithms— Reflections Based on the Book's Design

Meike Hardt

From the moment I was asked to do the design concept, I was excited about exploring the interrelations between humans and robots playfully and metaphorically as a design principle. Here, I understand the term robot in its broadest sense and use it as an analogy for software actors which, to a certain extent, act on their own initiative—and thus interfere with or have an influence on the design. For the layout of this book the "robots" in question are InDesign scripts for randomizing font selection and picture placement, and Python/InDesign scripts for generating background colors based on sentiment analysis.

I'm a German multidisciplinary designer working at the intersection of design and design research, interested in the political dimensions of design. I consider the materialization of the book and its mode of production to be as much a form of knowledge production as the contents. Language is not neutral, and neither is its visualization or representation and their modes of production. Understanding designing as a form of thinking through making, I want to share some brief and unfinished thoughts that came up while playfully exploring this human-robotic collaboration. I wondered whether there were any preferred, unintended, or even problematic design moments from which we can learn.

You have surely noticed the terms human and robot highlighted throughout the publication using different fonts. This is the first of the script-driven design interventions, all built together with Jef Van den broeck, a Belgian programmer. Our intention is to render the interdependencies between humans and robots visible by pointing out the various types and definitions of actors within the robotic research discourse. For this typographic intervention, we intentionally chose fonts with different design approaches, each of them carrying their own social, political and techno-historic story: Firstly, *Maison Extended* is a grotesque typeface that pays attention to harmony, rhythm, and flow—a font representing parameters often used for universal design approaches. Secondly, and contrary to this, *Serifbabe* by Charlotte Rhode takes a feminist perspective on type design, understanding font design symbolically as visual voices. And lastly, *Metafont*, a computer language for parametric type design that describes pen strokes of a letterform with mathemathical equations to provide a variety of font sets, plays an important role in computer and type history. Of course, these different positions represent only a small fraction of the diverse stories that exist in the type design world. Knowing all too well that many narratives are still not yet told and represented enough, questioning which (and in what ratio) social, political, and techno-historical knowledge we publish is relevant (Förster and Hardt, 2022). We fed the script with this list of fonts and let it randomize their replacements to see how the script renders these different design approaches visible. Since every single step needs to be taught to the script, handing over responsibility for how and what content is juxtaposed is only possible to a limited extent. Ultimately, such a juxtaposition was our responsibility rather than the script's, which requires reflection on the knowledge the script needs to be fed with.

Another design intervention using algorithms is the definition of the background color for the contributions under the heading *Perspectives*. This is intended to be a playful experiment about how color definition could be performed by an algorithm. The colored backgrounds are defined by sentiment analysis. The more positive the text, the lighter the background; the more negative the text, the darker. Such sentiment analysis is often used for evaluating product reviews and social media posts, so of course there is no claim to accurate analysis of this book's far more complex content. Sentiment analysis uses a list of known adjectives, each scored on its positivity and objectivity, to infer the sentiment of a given sentence. What scores should we assign to the term robotic, I wondered—neutral and objective? Decisions made on how words are scored on its positivity and objectivity are usually based on existing data sets. The problem with data sets is that they refer to only a moment in time, a certain cultural setting, and thus reproduce only a subset of social realities (Lücking, 2020). Playing with sentiment analysis to define background colors for this book raised the following questions: To what extent can an algorithm really capture all the sentiments contained in a text and are they visually represented as the authors intended? Is the algorithm not lacking the necessary experiential knowledge to ably conclude what counts as positive or negative, and what biases might such a word list contain? Here, again, reflecting upon what kind of knowledge the script needs to be fed with and from which perspective such a list needs to be designed are relevant to look at.

Indeed, this brief commentary does not provide space for in-depth analysis. Algorithms as being hidden and hard to grasp,

dynamics of knowledge power structures that algorithmic systems can (re-)produce certainly need to be explored in more detail. Knowing all too well that visual formal languages do not change the world, surely reflections on algorithms need to happen even beyond designing a book.

One inspiring way for me to think about knowledge production is the poetic and philosophical concept of "Sensuous Knowledge" by the Finnish-Nigerian feminist author and journalist Minna Salami (2020). From an African-centric and feminist perspectives, Salami proposes *Sensuous Knowledge* as a concept of combining emotional intelligence with interlectual intelligence. Interlectual intelligence refers to the knowledge of the mind, such as data, facts, reasoning and analysis; and emotional intelligence refers to the knowledge of the gut, such as experiential knowledge. The latter often takes a lower priority in knowledge production across Western society, leading to the exclusionary mechanisms we encounter in this world. With *Sensuous Knowledge* she proposes to understand knowledge production as a kaleidoscopic interweaving of both interlectual and emotional intelligence—without inherent binarism—to reveal *Europatriachal knowledge* and its inherent powerstructures. Such a kaleidoscopic understanding of knowledge production could help create awareness of the knowledge and power structures produced when designing with algorithms. Thus, how can this philosophical and poetic approach *Sensuous Knowledge* inform the practice of designing with algorithms, and what would a kaleidoscopic way of designing look like? The design of this book with algorithms represents a first attempt to explore these questions.

Salami, M. (2020). *Sensuous Knowledge: A Black Feminist Approach for Everyone.* Bloomsbury Publishing.

Förster, M. & Hardt, M. (2022). Critical by design? The book's design as SF figures. In: Mareis, C; Greiner-Petter, M; Renner, M. (2022). *Critical By Design? Genealogies, Practices, Positions.* Transcript.

Lücking, P. (2020). Automatisiere Ungleichheit. Wie Algorithmische Entscheidungssystem Gesellschaftliche Machtverhältnisse (re)produzieren. In netzforma* e.V. (Hrsg.), *Wenn KI dann feministisch. Impulse aus Wissenschaft und Aktivismus.* Netzforma* e.V.-Verein für feministische Netzpolitik, Berlin

# EDITORS, AUTHORS AND INTERVIEWERS AT THE GINA PROJECT

**FELIX CARROS** is a Researcher for the *Institute for Information Systems and New Media* at the *University of Siegen*. His research interest lies in applications for social-interactive robotics. He is investigating how it can be used in the context of banking, religion, and care through cross-cultural comparisons between Japan and Europe. He also researches participatory design in robotics.

**LARA CHRISTOFORAKOS** is a Graduate Research Assistant at the Chair for *Economic and Organizational Psychology* at *Ludwig-Maximilians-Universität München* (LMU, Munich, Germany) with a focus on the field of HCI. Her research explores psychological mechanisms and design factors in the context of technology use within various fields, such as companion technologies or social robots.

**DR. SARAH DIEFENBACH** is a Professor of Market and Consumer Psychology at *Ludwig-Maximilians-Universität München* (LMU, Munich, Germany) with a focus on the field of interactive technology. Her research group explores design factors and relevant psychological mechanisms within the context of technology usage across different fields (e.g., social media, digital collaboration, companion technologies, and social robots).

**DR. JUDITH DÖRRENBÄCHER (ED.)** is a Design Researcher at the Chair of *Ubiquitous Design/Experience and Interaction* at the *University of Siegen*. Educated in design, her current focus is on performative methods in design, theories about animism transferred to human-computer interaction (HCI) and design (techno-animism), and the interactions and design strategies of social robots.

**JOCHEN FEITSCH** is Researcher, Laboratory Manager, Project Coordinator, and Developer at the *University of Applied Sciences Düsseldorf*. On the Mixed Reality and Visualization team, he works in the field of human-technology interaction with a focus on interactive installations, experience design, motion capturing, creative coding, and augmented humans.

**DR. MARC HASSENZAHL (ED.)** is Professor of *Ubiquitous Design/Experience and Interaction* at the *University of Siegen*. He combines his training in psychology with a love for interaction design. With his group of designers and psychologists, he explores the theory and practice of designing pleasurable, meaningful, and transformational interactive technologies.

**STEPHANIE HÄUSLER-WEISS** is the Senior User Experience Designer at *User Interface Design GmbH*. Her work includes gathering and evaluating user requirements, elaborating interaction concepts, and realizing conceptual ideas graphically in screen and icon designs with all necessary controls.

**DR.-ING. THOMAS HULIN** heads the *Telemanipulation Research Group* at the *Institute of Robotics and Mechatronics of the German Aerospace Center (DLR)*. His research interests include haptics (control, devices, and algorithms), telerobotics, physical human-robot interaction, robot visualization, augmented reality, and skill transfer.

**JOHANNA LANGENDORF** studies HCI at the *University of Siegen* and works in the field of the use of robots in nursing. She has a bachelor's degree in Communication Psychology.

**DR. CATRIN MISSELHORN** is a Professor of Philosophy at the *University of Göttingen*. Her research focuses on the philosophical problems of artificial intelligence (AI), robot and machine ethics, and the ethical aspects of human-machine interaction. She has led a number of third party-funded projects particularly for the ethical evaluation of assistive systems and robots in care, the workplace, education, and other settings.

**ROBIN NEUHAUS (ED.)** is a Research Assistant at the Chair for *Ubiquitous Design/Experience and Interaction* at the *University of Siegen*. With a background in industrial design and HCI, his current research focuses on the design of experiences and interactions with robots, voice assistants, and other non-human actors.

**ADRIAN PREUSSNER** is a Research Assistant at the *Institute for Information Systems and New Media* at the *University of Siegen*. With a background in mechatronics, his research interests focus on HCI, the User Experience (UX) for AI (including Augmented Reality, AR), voice assistants, and the application of social interactive robotics in care settings.

**DAVE RANDALL** is Senior Professor at the *University of Siegen*. His interests are mainly having to do with computer-supported cooperative work (CSCW) and human-computer interaction (HCI) and the use of ethnographic methods for design purposes. He has co-authored five monographs and edited three books.

**RONDA RINGFORT-FELNER (ED.)** is a Research Assistant at the Chair for *Ubiquitous Design/Experience and Interaction* at the *University of Siegen*. With a background in design and HCI, her research focuses on design fiction, the design and exploration of future intelligent autonomous systems (such as social robots), and the exploration of related societal and social implications.

**KILIAN RÖHM** is a User Experience Consultant at *User Interface Design GmbH*. He designs and conceptualizes digital systems and services from a user perspective. His research interests include the design and exploration of intuitive, user-centered, and persuasive human-robot interaction.

**MANUEL SCHEIDEGGER** studied Philosophy and Dramatic Arts at *Freie Universität Berlin* and *Universität Hildesheim*. As a facilitator, he stages events and workshops on current topics (such as AI, new work, sustainability, and diversity) for organizations and the public; see www.argumentedreality.de.

**LISA SCHIFFER** joined *LAVAlabs* in 2018 as Project Manager and Concept Developer for research and development (R&D) projects after working at *Film- und Medienstiftung NRW* and the *Goethe Institut Santiago de Chile*. For *LAVAlabs*, she works in the field of interactive applications in human-technology interaction, incorporating virtual, augmented, and mixed reality.

**DR. TOBIAS STÖRZINGER** is a Researcher at Chair of Prof. Catrin Misselhorn (*University of Göttingen*). His work lies at the intersection of social ontology, ethics, and digital technology ranging from AI and robotics to Big Data.

**DANIEL ULRICH** is a Researcher at the *Institute of Informatics* at *Ludwig-Maximilians-Universität München* (LMU, Munich, Germany). His research focuses on interactions with (and the influence of) robots in the field of human-robot interaction, particularly robot personality and applications for social psychological mechanisms.

**DR. BERNHARD WEBER** is Human Factors Expert of the *Institute of Robotics and Mechatronics* at the *German Aerospace Center* (DLR). His research focuses on human-centered development and the evaluation of telerobotic systems (e.g., in the field of surgery, nursing, and space robotics) and haptic interaction technologies.

**DR. RAINER WIECHING** is Senior Research Scientist at the *University of Siegen*. He has successfully coordinated several health- and robotics-related national and international R&D projects (H2020, BMBF, DAAD). These projects centered on designing meaningful socio-technology support systems for active and healthy aging, preventive medicine, and care robotics.

**ANNE WIERLING** studied at the *University of Applied Science in Osnabrück* and graduated with a Master of Law in 2018. Since January 2019 she is a researcher at the chair of Prof. Dr. Becker (since 2020 at the chair of Prof. Dr. Krebs) at the *University of Siegen* for the research project GINA. Her research interest lies in data protection combined with AI in the health area.

**VOLKER WULF** is Professor of *Information Systems and New Media* at the *University of Siegen*. His research interests lie primarily in the area of socio-informatics, taking a practice-based approach to the design of information technology (IT) systems in real-world settings.

# INTERVIEW PARTICIPANTS IN EIGHT ROBOTICS PROJECTS

## ERIK

**SIMONE KIRST** is a Research Associate in the *Clinical Psychology of Social Interaction Group* at *Humboldt Universität zu Berlin*, Germany, where she focuses on technology-based clinical interventions for autistic children (for example, see www. zirkus-empathico.de). For *ERIK*, she conceptualized human-robot interaction strategies and evaluated the robotic system.

**MILENKO SAPONJA** is an Affective Robot Researcher at *audEERING GmbH*, a company that provides intelligent audio solutions. For *ERIK*, he worked on acoustic emotion recognition and its implementation within the robotic system.

**JULIAN SESSNER** is a Research Associate at the *Institute for Factory Automation and Production Systems* at *Friedrich-Alexander-Universität Erlangen-Nürnberg*. The focus of his research is assistance systems and service robots in medical scenarios. For *ERIK*, he worked on the development of a technical infrastructure for robot-assisted autism therapy.

**MARTINA SIMON** is a Research Associate and Project Manager in the *Human Centered Innovation Group* at the *Fraunhofer Center for Applied Research on Supply Chain Services* at *Fraunhofer IIS*. Her research focuses on topics in human-technology interaction as well as technology/AI acceptance and use.

**MARTIN STREHLER** is one of the founders of *Innovationsmanufaktur*, where he has been moderating innovation projects for over 20 years. He is an expert in innovation methods and involving users in innovation processes. For *ERIK*, he designed the study of interaction between children and the robot Pepper.

## INTUITIV

**DIPL.-ING. KARSTEN BOHLMANN** is head of Research & Development at *ek robotics GmbH*. His work focuses on the transformation of automated guided vehicles towards autonomous transport robots.

**HANNS-PETER HORN** is a psychology graduate in Engineering Psychology and Human Factors. He graduated from *Humboldt University*, Berlin, and has since worked as an expert in general UX, human-robotic-interaction (HRI), and psychological methodology/ statistics at *HFC Human-Factors-Consult GmbH* in Berlin.

**DR.-ING. TIM SCHWARTZ** is Senior Researcher at the *German Research Center for Artificial Intelligence* (DFKI) in Saarbrücken, Germany. He is head of the *Human-Robot Communications Group* and the *German-Czech Innovation Lab for* Human-Robot *Collaboration in Industrie 4.0 (MRK 4.0)*.

## I-RobEka

**ANDY BÖRNER** is a Research Associate at *Chemnitz University of Technology*. After studying Media Informatics and Computer Science, he worked in the areas of human-robot interaction and participatory user studies for the *I-RobEka* project. His other research interests include data analysis and the ethical, social, and legal aspects of technology.

**DR. GUIDO BRUNNETT** is Full Professor of *Computer Graphics and Visualization* and head of the competence center *Virtual* Humans at *Chemnitz University of Technology*. His research focuses on the creation of digital humans and their use in VR applications. Currently, he is especially interested in the synthesis of human motions.

**JAN LINGENBRINCK** grew up on the Dutch border and has been working at the intersection of retail and tech since 2003 for global retailers. He is currently scouting relevant technologies for *EDEKA Digital* at *Food Tech Campus* in Berlin. He admires the entrepreneurial mindset of *EDEKA*'s independent retailers. He has two children and loves to eat, ride motorcycles, and play video games.

## KoBo34

**CAREN HORSTMANNSHOFF** is Research Associate at *Rosenheim Technical University of Applied Sciences* assigned to conduct the evaluation for *KoBo34*. As a physiotherapist and health scientist, her research interests are complex interventions to obtain or improve mobility and participation in older people.

**EVA THERESA JAHN** is a Communication Scientist and Research Assistant at *The Munich Institute of Robotics and Machine Intelligence* (MIRMI) at the *Technical University of Munich*. For *KoBo34*, she worked on the evaluation work package. Her research focuses on acceptance studies in the field of HRI and the relationship between robot design and sympathy.

**CHRISTOPH JÄHNE** studied Mechanical Engineering at *TU München* from 2008–2014. From 2015–2017, he worked as a research associate at *TUM-ITR* (Prof. Dr. Sandra Hirche). Since 2017, he has worked as a senior developer at *Franka Emika* (R&D, service robotics).

**DOROTHEA KOERT** is currently an Independent Research Group Leader at *TU Darmstadt*, where she works on interactive AI and human-robot interaction. For *KoBo34*, she worked on the skill-learning work package; her research focused on robot skill learning from users with little or no prior experience with robotics.

**MARTIN MÜLLER** is Professor and Nurse Researcher at *Rosenheim Technical University of Applied Sciences*. For *KoBo34*, he led the evaluation work package. His main research interests are the development and evaluation of complex nursing interventions to improve participation among older people in nursing homes and transitional care.

**SUSANNE TRICK** is a PhD student at the *Psychology of Information Processing Laboratory* at *TU Darmstadt* and part of the IKIDA research group investigating interactive AI algorithms. Her research focuses on the understanding and prediction of human behavior, particularly in human-robot interaction. To that end, she also works on the optimal combination of multimodal data.

# MIRobO

**KAI-UWE KADEN** graduated with a degree in engineering from the *University of Applied Sciences in Mittweida*. Following his studies, he worked for over a decade as an electronics engineering and manufacturing services provider in various roles, including project and program manager. Since June 2018, he has headed the Smart Systems Division at *FusionSystems GmbH* in Chemnitz. In this role, he is in charge of R&D projects with a particular focus on recognition functions using AI/machine-learning methods.

**DOROTHEA LANGER** is a graduate in Psychology of *Chemnitz University of Technology*. Since 2015, she has continued her education as a Research Assistant at the Chair for *Ergonomics and Innovation*. Her research is focused on perception in virtual 3D environments and its usage for evaluation in ergonomics, as well as on issues of human-robot collaboration.

**DANIEL-PERCY WIMPFF** is a state-certified Computer Scientist of the *Academy for Data Processing (ADV)* in Böblingen. He has been working as a Software Engineer since 1999 including research and development and project management for various collaborative research projects from 2007 on. Since December 2014 he has been a Software Developer at *Sikom Software GmbH* in Stuttgart, concerned with voice and sound recognition and call center software.

# NIKA

**KATHRIN POLLMANN** is a User Experience Researcher at the *Fraunhofer Institute for Industrial Engineering IAO*. In her research, Kathrin adopts a human-centered, need-based design approach with the goal of creating positive experiences with technology. Her current work focuses on the design of social human-robot interaction.

**SABINE SCHACHT** is a Catholic Theologian and Researcher and Doctoral Candidate in the *International Centre for Ethics* (IZEW) at *University Tübingen*. Her research focuses on the ethics of care and caring communities in relation to palliative care questions. She also works in the field of robotics in (elder) care and since 2015 she has been working in the field of terminal care at the *Tübinger Hospizdienste e.V.*

**DANIEL ZIEGLER** is a Computer Scientist and Researcher at *Fraunhofer IAO* in the area of user experiences and the human-centered design of interactive systems. In his current research, Daniel focuses on the automatic personalization of user interfaces and human-robot interaction in order to address the individual needs, abilities, and preferences of a diverse range of users.

## RobotKoop

**FRANZISKA BABEL** is a PhD student in the *Department of* Human *Factors* at *Ulm University* in Germany. She received a Master of Science from *Ulm University* in 2018. Her research interests include human-robot interaction, persuasive robotics, and human-machine cooperation.

**DR. SIEGFRIED HOCHDORFER** is the Co-Founder and CTO of *ADLATUS Robotics*. He studied Computer Engineering and Applied Computer Science. During his research activities, he published more than 20 scientific papers, mainly on localization and navigation of autonomous mobile systems. He also lectures on Autonomous Mobile Systems at *Ulm University of Applied Science*.

## VIVA

**SONJA STANGE** is a PhD student in the *VIVA* project from the *Social Cognitive Systems Group* at *Bielefeld University*. Her research interest lies in equipping a lively social robot with the ability to provide behavioral explanations in an intuitively understandable way. She investigates the effects of such explanations on user perception of the robot and its behavior.

**JULIA STAPELS** is a PhD student in the *VIVA* project from the *Applied Social Psychology and Gender Research Group* (Prof. Dr. Friederike Eyssel) at *Bielefeld University*. She is interested in improving measurement validity concerning attitudes towards robots as well as investigating the causes and consequences of robot-related attitudinal ambivalence.

**CLAUDE TOUSSAINT** studied Mechanical Engineering and Product Design. As the owner of a design agency with 100 employees, he has advised renowned international brands on user experience design and product strategy. In 2017, he founded the start-up *Navel Robotics*, which is developing a social robot figure called *Navel* for marketing in 2023.

# CONTRIBUTORS OF THE "PERSPECTIVES"

**JAMES AUGER** is an Enseignant Chercheur and Directeur of the department of design at the *École Normale Supérieure Paris-Saclay*. His practice-based design research examines the social, cultural and personal impacts of technology and the products that exist as a result of development and application.

**JOHANNA BENZ** founded *Graphic recording.cool* in 2013 together with Tiziana Beck. Based in Leipzig, Berlin and Paris, the duo develops individual graphic recording formats. Through the live-drawing process, they illustrate and comment on facts and ideas and show subjective images with the aim of creating new connections between visual and applied arts, science and education.

**ELKE BUTTGEREIT** is a Vice-Principal at an elementary school in Bonn. She studied to become a Primary School Teacher in Cologne, specializing in German, math and general studies (social studies). She has worked as a Primary School Teacher for more than 30 years. She has been grappling with the importance of robotics in schools since 2018.

**BRIGITTA HABERLAND** is a Pastoral Counselor at the *Protestant University in Rhineland-Westphalia-Lippe*, and a Psychosocial Counselor at the *University of Health in Bochum*. After studying Protestant theology and social pedagogy, she worked as a Project Manager in the field of school and social inclusion of people with disabilities at a church agency.

**MEIKE HARDT** is a Designer and Design Researcher interested in the political dimension of design. She is co-founder of the *Research Institute of Botanical Linguistics (RIBL)* and worked for the project *Critical by Design?* at the *Institute of Experimental Design and Media Cultures (IXDM), FHNW Basel* as a researcher and coordinator.

**DR. MARC HASSENZAHL** is Professor of *Ubiquitous Design/Experience and Interaction* at the *University of Siegen*. He combines his training in psychology with a love for interaction design. With his group of designers and psychologists, he explores the theory and practice of designing pleasurable, meaningful, and transformational interactive technologies.

**EDI HAUG** lives near Berlin and catches most of his Pokémon in Prague, London and the south of Berlin, with the help of his human friend, Laura, or his family. Due to a hereditary disease, he has lived deaf and blind since he was 12 years old. He is currently studying shorthand braille and SMS writing, and he wants to become a masseur.

**ANTJE HERDEN** has written popular books for children and young adults for publishing houses like *Tulipan*, *Fischer*, and *Beltz*. She was awarded the Peter Härtling Prize in 2019 for *Keine halbe Sachen* and was nominated for the German Children's Literature Award. She lives in Darmstadt and has two children.

**DR. TIMO KAERLEIN** is Akademischer Rat at the Institute for Media Studies at *Ruhr University Bochum*. His work focuses on the theory, history, and aesthetics of interfaces, social robotics, embodiment relations of digital technologies, and approaches to urban affective sensing.

**DR. LENNEKE KUIJER** is Assistant Professor in the *Future Everyday Group* in the *Department of Industrial Design* at the *TU Eindhoven* (NL). Her research combines sociology and design theory to better understand the role of interactive technologies and their designers in the dynamics of everyday life, particularly around energy demand in the home.

**DR. JANINA LOH** is an ethicist (Stabsstelle Ethik) at *Stiftung Liebenau* in Meckenbeuren on Lake Constance. They received their doctorate at the *Humboldt University* in Berlin. Their narrower research interests include responsibility, trans- and posthumanism, and robot ethics, as well as Hannah Arendt, feminist philosophy of technology, theories of judgment, and ethics in the sciences.

**DR. ILONA NORD** is Professor of Religious Education and Practical Theology at the *University of Würzburg*, Germany. Her research focuses on religion and media, digital literacy, and, digital religion. She also works on diversity management in cultural contexts, antisemitism, and religious pluralism. Her latest publications are the compendium *Theologie und Digitalität* (Herder 2021, co-edited with W. Beck & J. Valentin) and *Churches Online in Times of Corona (CONTOC): First Results* (Campbell 2021, with O. Adam).

**DR. CORINNA NORRICK-RÜHL** is Professor of Book Studies in the *English Department* at the *University of Muenster*, Germany. In her research and teaching, she focuses on 20th- and 21st-century publishing history and book culture. Recent publications include *Book Clubs and Book Commerce* (Cambridge University Press, 2020) and the volume *The Novel as Network: Forms, Ideas, Commodities* (Palgrave, 2020, co-edited with T. Lanzendörfer).

**UWE POST** is a Software and Game Developer, IT Consultant and Lecturer, as well as the Author of IT reference books and mostly satirical science fiction novels. *Walpar Tonnraffir und der Zeigefinger Gottes* was awarded the German Science Fiction Prize in 2011, as well as the Kurd Laßwitz Prize. Post lives with his family on the southern edge of the Ruhr area in Germany. www.uwepost.de

**SCARLET SCHAFFRATH (SIEBERT)** is a Communication Scientist and PhD student at *TH Köln—University of Applied Sciences,* Institute for Media Research and Media Education, and analyzes future visions of social robots in educational contexts. In her research she focuses on participatory approaches to technical innovation. She also works as a scientific consultant at *VDI/VDE-IT GmbH.*

**LAURA M. SCHWENGBER** is a Sign Language Interpreter, has an MA in barrier-free communication and is writing her PhD at the *Institute for Easy and Plain Language* at *University of Hildesheim.* Edi Haug has been her best friend for 20 years now. She lives in Berlin and teaches people how to write barrier-free social media postings with her team #BarrierefreiPosten.

**DR. KARSTEN WENDLAND** is Professor of Media Informatics at *Aalen University* and Senior Researcher at the *KIT.* He studied computer science at *Darmstadt University of Technology*, was a research associate at the Center for Interdisciplinary Technology Research (ZIT), and earned his doctorate in technology design.